# Springer Theses

## Recognizing Outstanding Ph.D. Research

## Aims and Scope

The series "Springer Theses" brings together a selection of the very best Ph.D. theses from around the world and across the physical sciences. Nominated and endorsed by two recognized specialists, each published volume has been selected for its scientific excellence and the high impact of its contents for the pertinent field of research. For greater accessibility to non-specialists, the published versions include an extended introduction, as well as a foreword by the student's supervisor explaining the special relevance of the work for the field. As a whole, the series will provide a valuable resource both for newcomers to the research fields described, and for other scientists seeking detailed background information on special questions. Finally, it provides an accredited documentation of the valuable contributions made by today's younger generation of scientists.

## Theses are accepted into the series by invited nomination only and must fulfill all of the following criteria

- They must be written in good English.
- The topic should fall within the confines of Chemistry, Physics, Earth Sciences, Engineering and related interdisciplinary fields such as Materials, Nanoscience, Chemical Engineering, Complex Systems and Biophysics.
- The work reported in the thesis must represent a significant scientific advance.
- If the thesis includes previously published material, permission to reproduce this must be gained from the respective copyright holder.
- They must have been examined and passed during the 12 months prior to nomination.
- Each thesis should include a foreword by the supervisor outlining the significance of its content.
- The theses should have a clearly defined structure including an introduction accessible to scientists not expert in that particular field.

More information about this series at http://www.springer.com/series/8790

Liqiang Xu

# Impact of Climate Change and Human Activity on the Eco-environment

## An Analysis of the Xisha Islands

Doctoral Thesis accepted by
the University of Science and Technology of China,
Hefei, People's Republic of China

Springer

*Author*
Dr. Liqiang Xu
School of Resources and Environmental Engineering
Hefei University of Technology
Hefei
People's Republic of China

*Supervisors*
Prof. Liguang Sun
Dr. Xiaodong Liu
Institute of Polar Environment
University of Science and Technology of China
Hefei
People's Republic of China

ISSN 2190-5053 ISSN 2190-5061 (electronic)
ISBN 978-3-662-45002-4 ISBN 978-3-662-45003-1 (eBook)
DOI 10.1007/978-3-662-45003-1

Library of Congress Control Number: 2014950522

Springer Heidelberg New York Dordrecht London



Printed on acid-free paper

Springer is part of Springer Science+Business Media (www.springer.com)

**Parts of this thesis have been published in the following journal articles:**

1. **Xu L.Q.**, Liu X.D.*, Jiang S. Late-Holocene seabird palaeodietary record from Ganquan Island, South China Sea. Quaternary International 333: 139–145, 2014. [Reproduced with permission from Elsevier]
2. **Xu L.Q.**, Liu X.D.*, Sun L.G.*, Liu W.Q., 2012. Rapid identification of source materials in coral sand ornithogenic sediments by reflectance spectroscopy. Ecological Indicators 23: 517–523, 2012. [Reproduced with permission from Elsevier]
3. **Xu L.Q.**, Liu X.D.*, Sun L.G*, Chen Q.Q., Yan H., Liu Y., Luo Y.H., Huang J. A 700-year record of mercury in avian eggshells of Guangjin Island, South China Sea. Environmental Pollution 159: 889–896, 2011. [Reproduced with permission from Elsevier]
4. **Xu L.Q.**, Liu X.D.*, Sun L.G.*, Yan H., Liu Y., Luo Y.H., Huang J. Geochemical evidence for the development of coral island ecosystem in the Xisha Archipelago of South China Sea from four ornithogenic sediment profiles. Chemical Geology 286: 135–145, 2011. [Reproduced with permission from Elsevier]
5. **Xu L.Q.**, Liu X.D.*, Sun L.G.*, Yan H., Liu Y., Luo Y.H., Huang J. A 2200-year record of seabird population on Ganquan Island, South China Sea. Acta Geologica Sinica-English Edition 85(4): 957–967, 2011. [Reproduced with permission from John Wiley and Sons]
6. **Xu L.Q.**, Liu X.D.*, Sun L.G.*, Yan H., Liu Y., Luo Y.H., Huang J., Wang Y.H. Distribution of radionuclides in the guano sediments of Xisha Islands, South China Sea and its implication. Journal of Environmental Radioactivity 101: 362–368, 2010. [Reproduced with permission from Elsevier]
7. Liu X.D.*, **Xu L.Q.**, Chen Q.Q., Sun L.G.*, Wang Y.H., Yan H., Liu Y., Luo Y.H., Huang J. Historical change of mercury pollution in remote Yongle archipelago, South China Sea. Chemosphere 87: 549–556, 2012. [Reproduced with permission from Elsevier]
8. Liu X.D.*, **Xu L.Q.**, Sun L.G.*, Liu F., Wang Y.H., Yan H., Liu Y., Luo Y.H., Huang J. A 400-year record of black carbon flux in the Xisha archipelago, South China Sea and its implications. Marine Pollution Bulletin 62: 2205–2212, 2011. [Reproduced with permission from Elsevier]

* Corresponding author

# Supervisors' Foreword

On 27 September 2013, Working Group I of Intergovernmental Panel on Climate Change (IPCC) published a report namely "Climate Change 2013: The Physical Science Basis, Summary for Policymakers." According to this report, multiple independent proxy data showed that Earth surface temperature increased by 0.85 °C (0.65–1.06 °C) over the period 1880–2012. Moreover, the past three decades (1983–2012) are likely the warmest period over the past 1,400 years. The warming seems worse than we thought in 2007 when IPCC released its fourth assessment report (AR4). Whatever the reason is, global warming is an inconvenient fact and has drawn increasing attention from scientific groups, as well as political organizations. For example, the 2007 Nobel Peace Prize was awarded to former vice president of the United States Al Gore and IPCC.

As human beings, we are more concerned about the future of the Earth under successive warming. The question is what will happen to ecosystems on the Earth, and whether lives can bear such a rapid temperature change. To a geologist, it is a basic principle that the past is the key to the future. In other words, to well predict the future, we need to understand climate change in the past and its impacts on ecosystems. Although we cannot go back to the past by something like time-travel, natural archives can tell us the history of the Earth. Scientists have obtained a large number of reliable climatic records at different timescales by myriad archives, e.g., ice cores, loess, deep-sea sediments, stalagmite, coral and tree rings, etc. The responses of earth ecosystems to climate change, however, have been less documented. Our research group has been working on the topic of ecological responses to climate changes over the last decade.

In addition to abrupt climate change, influences of human activities nowadays on the environment has been unprecedented, and also imposed significant impacts on the environment. In the year 2013, a new scientific journal "Anthropocene" was published to address the extent of interactions that human have with Earth. We also would like to identify any evidence of past human activity and examine its possible impacts on the environment.

We have studied ecological changes in birds and mammals, for example penguins and seals, within the past 3,000 years in high-latitude Antarctica and Arctic.

Eleven years ago, we began to study a new low-latitude area, the Xisha Islands in the central South China Sea. The evolution of the ecosystem on these coral Islands was developed, which is very different from that of the Antarctica and Arctic. A study on the island ecosystem is a new challenge for us. The Xisha Islands have been a world-famous seabird habitat, but birds have almost completely disappeared today. Based on our earlier research, Dr. Xu started his Ph.D. study. It is expected to find the answer or some clues in this Ph.D. thesis.

Using bird-affected sediments (ornithogenic sediments) and avian subfossils, Dr. Xu reconstructed seabird ecology, including seabird population and dietary changes, over the last 2,000 years in this Ph.D. thesis. The changes in seabird ecology in response to climate change and anthropogenic activity is discussed in-depth. It is found that both are partly responsible for changes in seabird ecology. This work provides a scientific basis for future ecological study and resource exploitation on these coral islands.

Hefei, August 2014

Prof. Liguang Sun
Dr. Xiaodong Liu

# Abstract

Global warming significantly impacts on ecosystems and has drawn increasing attention in recent years. However, how climate changes affect lives and the environment remains unclear. Seabirds move across aquatic/terrestrial ecosystem boundaries and their fossil remains have been widely used for studying ecological responses to climate change. Five sediment cores from Ganquan (GQ), Guangjin (GJ3), Jinqing (JQ), Jinyin (JY2), and Chenhang (CH) islands of the Xisha archipelago, South China Sea, were collected during field investigations. Well-preserved seabird remains were observed in these sediment cores.

Via analysis of several natural and anthropogenic radionuclides in the ornithogenic sediments, $^{210}Pb$ dating has been proven to be effective for coral sand sediments in the Xisha Islands. The average supply rate of $^{210}Pb$ was 126 Bq $m^{-2}$ $a^{-1}$, very close to the flux of the northern hemisphere average (125 Bq $m^{-2}$ $a^{-1}$). Radiocarbon analysis of ancient bones showed that the most ancient sample of these cores dated back to approximately 2000 years. Geochemical characteristics of four profiles GQ, GJ3, JQ, and JY2 were analyzed. The results suggest that the source materials of ornithogenic sediments have changed gradually from a two-component (coral sand, guano) mixture to a three-component (coral sand, guano, and humus) mixture, likely indicating the slow development of vegetation following seabird occupation. Cu, Cd, Zn, P, As, Se, and Ba were identified as a group of avian bio-elements. A 2200-year record of seabird population on Ganquan Island was further reconstructed by avian-bioelements and reflectance spectroscopy. A cool climate during the Little Ice Age (LIA) seems more favorable to seabirds on the Xisha Islands. Relative low sea surface temperature (SST), stronger monsoons, and weaker ENSO activity during the LIA might result in increase in marine nutrient mixing and high primary productivity. This provided high food availability and thus supported more seabirds. Moreover, a change in nitrogen/carbon isotope compositions in the collagen of seabird bones suggested that seabirds might dive deeper or travel further to catch prey, due to intra/inter-competitions. The abrupt decline in seabird population in recent times is probably attributed to human activity. The level of Hg in eggshells and the bulk ornithogenic sediments was determined. Eggshell Hg and Hg flux well-recorded past human activity. We reconstructed a 400-year record of black carbon (BC) deposition flux from profiles GJ3, JQ, and JY2. In the recent 30 years, the BC flux displayed decreasing trend, very likely due to a change in energy structure and development of pollution control techniques. Preliminary analysis of DNA in ancient bird droppings suggests that the guano contains quite a low level of ancient DNA, due to possible severe degradation.

**Keywords** Climate change · Xisha Islands · Seabirds · Stable isotope · Ecological response · Human activity

# Acknowledgments

Foremost, I would like to express my deepest gratitude to Professor Sun Liguang (USTC), Dr. Liu Xiaodong (USTC), and Dr. Lam Yun Wah (CityU) for their patient guidance, encouragement, and financial support throughout my Ph.D. study. Dr. Liu has consumed much of his time and energy on my study since I came to the Institute of Polar Environment, University of Science and Technology of China several years ago. He supervised my first experiment, first illustration, and first paper, and helps me resolve any other difficulty that I encounter. I thank Dr. Lam for giving me the chance to study in Hong Kong and sharing his imaginative ideas with me. Moreover, Professor Cheng Shuk Han is acknowledged for her considerable concerns on my Ph.D. project and my study. Prof. Wang Yuhong from National Institute of Health, U.S.A. and Ms. Geraldine Carton from Soochow University are acknowledged for their considerable help in improving the English of our published papers. I would like to thank chemists Liang Youqing, Technician Liu Houqi, Dong Hongjuan, Eartha, Dr. Zhaoqing for their considerable technical support. I also learned analytical techniques from Wang Liangliang (Northeast Forestry University), Shi Ruiying (Soochow University) and Zhao Jiaohong (Shanghai Normal University).

I am quite lucky to have the chance to study in Hefei, Suzhou, and Hong Kong and I would like to show my gratitude to all my team members in Hefei, Suzhou, and Hong Kong. Members of Dr. Liu's research group, including Jiang Shan, Chen Qianqian, Nie Yaguang, Ma Xiaorong, Lou Chuangneng, etc., helped me quite a lot during the past several years.

Publication of this work was financially supported by the National Natural Science Foundation of China (Nos. 41402148, 41376124, 40730107), Knowledge Innovation Program of Chinese Academy Sciences (KZCX2-EW-QN50), China Postdoctoral Science Foundation (No. 2014M550338), and Fundamental Research Funds for the Central Universities (No. 2013bh2x0026). All members of the field study team, including Yan Hong, Liu Yi, Luo Yuhan, Huang Jing, and the Chinese People's Liberation Army, are acknowledged for their help in sample collection. Reviewers of our published papers are acknowledged for their constructive comments and suggestions, which improved the quality of our work. I also would like to thank editor PEI Xiaoli from Springer for her help in publication of this Ph.D. thesis.

Finally, I thank my families for their encouragement and support through all of my endeavors.

Liqiang Xu

# Contents

# Chapter 1
# Introduction

In the year 1962, American biologist Rachel Carson published her book "Silent Spring", which documented detrimental effects of pesticides on the environment. Since then, "environmental issue" appeared for the first time as a technical term. Of all typical environmental problems, global change has drawn increasing worldwide attention over the past decades. Climate change in the post-industrial era is unprecedented in human history. For example, the Earth surface temperature increased by approximately 0.85 °C over the period 1880–2012. The concentration of atmospheric $CO_2$ was above 400 parts per million (ppm) for the entire month of this April (2014). Such a $CO_2$ level deserves highest attention, as its concentration has never exceeded 300 ppm over the past 800,000 years.

Although the reasons for warming are sometimes debated, the theory of green house effect is overwhelming. Apart from its causes, global warming has been a painful fact we have to face. Large scale El Niño-Southern Oscillation (ENSO) and frequent extreme weather, e.g. floods, droughts and typhoons etc., are probably effects of global warming. So far, we know little about the impacts of global warming on organisms and the environments. The effects may be beyond our speculation and fatal to lives, including human being, on the Earth. Under a circumstance of successive warming, it is expected to know what will happen to creatures and ecosystems. This is of great importance for human social, economical and cultural development.

The Earth is a system that organisms interact with their inorganic surroundings. The science of Earth system studies interplay between the hydrosphere, atmosphere, lithosphere and biosphere. The present study is thus based on the rationale of Earth system science. Foremost, an ideal study area and some biological materials are required. The Xisha Islands in the central South China Sea is one of the perfect target areas, and the sediments influenced by seabirds serve as robust natural archives.



L. Xu, *Impact of Climate Change and Human Activity on the Eco-environment*, Springer Theses, DOI 10.1007/978-3-662-45003-1_1

Past is the key to the future. To answer the questions discussed above, we need to study ecological responses of ecosystems to climate and human activities in the past. The present research project is "Impact of Climate Change and Human Activity on the Eco-environment: An Analysis of the Xisha Islands". It is an interdisciplinary study, within the scope of ecological geology. Research associated with this study includes ecological geology, climate change over the last 2,000 years, biotransport, palaeoecology of seabirds, paleoceanography of the South China Sea, and ecology of coral islands. Advances in such study fields are demonstrated as follows.

## 1.1 Advances in the Study of Ecological Geology

Ecological geology is an interdiscipline that incorporates concepts and principles of geology, ecology, environmental science, archaeology, paleontology, palaeoclimatology etc. Its principal aim is to reconstruct eco-environment records by geological materials. Ecological geology studies environmental processes on various time-scales (from seasonal/annual to geologic). Nonetheless, most of ecological geology studies focus on a specific period, the Quaternary (from approximately 2.6 million years ago until now), during which climate was quite unstable (Liu 1997). Impacts of the environment on lives are reflected in a variety of aspects. Climate change can significantly influence macrostructures of organisms, e.g. population size, foraging behavior and reproductive strategy (Sun et al. 2000; Huang et al. 2011), as well as their DNA microstructures. For instance, Lambert et al. (2002) reported that the calculated evolution rates of Adélie penguins during the past 7,000 years were approximately two to seven times higher than previous phylogenetic estimates. According to mammalian diversity analysis, it was found that diversity of mammalian in North America over the Cenozoic was highly correlated with benthic foraminiferal oxygen isotope (global ice volume indicator), implying significant impacts of climate on diversity pattern of terrestrial mammals (Figueirido et al. 2012). Changes in marine environment can induce abrupt shift in seabird diets. Scientists investigated dietary record of penguin over the last 38,000 years and found an abrupt shift from fish to krill within the recent 200 years (Emslie and Patterson 2007). Professor Liu Tungsheng is one of the Chinese first generation ecological geologists. He devoted his whole life to the studies of Quaternary environment and Quaternary geology by unique Chinese loess, and estimated timing and duration of the Quaternary. His research group extended the onset of the Quaternary back to approximately 2.6 million years, making a great contribution to the study of ecological geology (Liu 1997). Using ornithogenic sediment archives from polar areas, professor Sun Liguang and his research group made significant progress in historical population reconstruction, environmental pollution, culture archaeology etc (Sun et al. 2000, 2006; Huang et al. 2011). By far, ecological geology remains at a preliminary stage; its rationale and methodology still need to be further improved.

Selection of suitable materials is a prerequisite in ecological geology study. Biological remains, such as ornithogenic sediments, furs/hairs, bones, fish scales etc., have been proven to be ideal materials for studying past environmental changes (Sun et al. 2000; Liu et al. 2006a; Huang et al. 2011; Emslie et al. 2007; Chamberlain et al. 2005; Lambert et al. 2002). Seabird droppings (guano) are waste products from their digestive tract, and provide valuable information of past environmental changes. Via elemental and isotopic analyses, they have been widely used to track changes in bird ecology (Emslie and Patterson 2007; Sun et al. 2000; Emslie et al. 2007; Lambert et al. 2002).

## 1.2 Climate Change Over the Past 2,000 Years

The records of climate over the past 2,000 years are essential for the present study. Studies have used various natural archives to reconstruct climatic processes during this period (Fig. 1.1, Mann et al. 2008). For example, Hu et al. (2001) identified climate oscillations over the last two millennia from analysis of lacustrine sediments in Alaska. Oppo et al. (2009) reviewed sea surface temperature (SST) and hydrological records during the past 2,000 years in the Indo-Pacific Warm Pool (IPWP). Utilizing marine sediments, Sicre et al. (2008) unveiled decadal variability of SST of North Iceland over this period. In South Africa, a 2000-year record of climate has also been well-reconstructed (Tyson and Lindesay 1992). On the basis of regional reconstruction, Michael Mann from University of Pennsylvania, USA and his colleagues collected a variety of proxy-based records worldwide, and

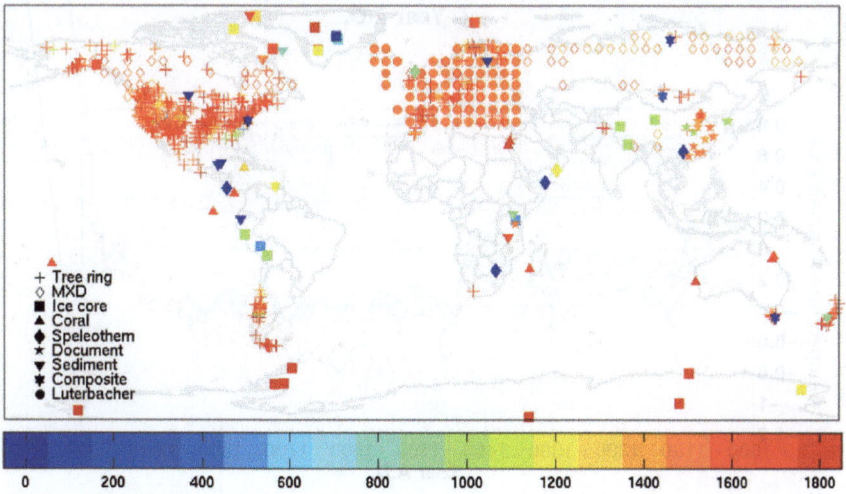


**Fig. 1.1** Materials used for reconstruction of climate over the last two millennia and their global distributions. [Reprinted with permission from Mann et al. (2008). Copyright (2008) National Academy of Science, USA]

discussed detailed characteristics of climate dynamics over the past 2,000 years (Mann et al. 1999; Mann and Jones 2003; Jones and Mann 2004). Temperature records of the Earth surface over the two millennia are shown in Fig. 1.2. National Research Council of the Academies, USA (2006) also released its synthesis report about global climate changes within the last two millennia. Critical climatic events, including Medieval Warm Period (MWP), Little Ice Age (LIA) and recent global warming, were specified in detail in this scientific report. The Intergovernmental Panel on Climate Change (IPCC) was established in 1988 by two organizations, the World Meteorological Organization (WMO) and the United Nations Environment Programme (UNEP) to solve international climatic crisis. Four scientific reports about earth climate have been released by IPCC ever since 1988. Its 4th synthesized report in 2007 revealed an evident increase

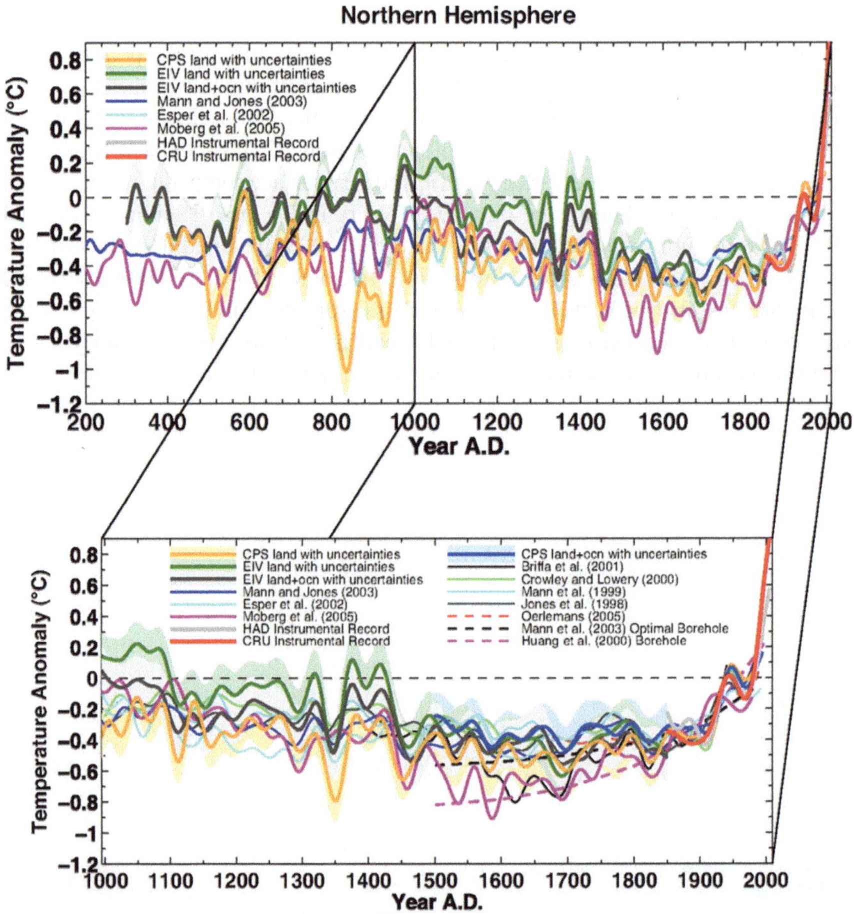


**Fig. 1.2** Proxy-based reconstruction of Northern Hemisphere temperature over the past 2000 years. [Reprinted with permission from Mann et al. (2008). Copyright (2008) National Academy of Science, USA]

in temperature by 0.74 °C over the past century, and it was very likely that the observed increase in earth temperature was predominantly due to anthropogenic forcing (IPCC 2007). IPCC also emphasized the necessity to promote study of ecological responses in its recent report "Intergovernmental Panel on Climate Change special report on managing the risks of extreme events and disasters to advance climate change adaptation".

It cannot be ignored that Chinese scientists also have made significant advances in the study of climate change over the past 2,000 years. For example, Zhang et al. (2008) identified potential relationships among climate, sun and culture from an 1810-year speleothem record. A 2650-year lamina record of stalagmite from Beijing unveiled unambiguous cyclic rapid climate oscillation on centennial scale (Tan et al. 2003). Thompson et al. (2000) analyzed Dasuopu ice core from the Qinghai-Tibet plateau, and they found evident changes of ice-core oxygen isotope compositions in response to rapid warming in the 20th century. Extracting information of past climate from historical documents is a superiority of Chinese scientists, as China has vast and high-resolution official history books. Geologist Chu Coching initiated the study of past climate by historical documents, and published his influential paper namely "A preliminary study on the climatic fluctuations during the last 5,000 years in China" (Chu 1972). Following Chu's study, Professor Zhang De'er from National Climate Center of China and her research group spent more than 20 years to write a monography "A Compendium of Chinese Meteorological Records of the Last 3,000 Years", reviewing the detailed meteorological events over the past several thousand years (Zhang 2004; Zhang and Lu 2007). This is a great contribution to the study of climate by historical documents. Chinese scientists have reconstructed meteorological records for different places in China via various natural materials, including sediments, ice cores, tree rings, historical documents, etc. Yang et al. (2002) supplied a new synthesized climate record by incorporating all these proxy-based records of climate over the last 2,000 years in China. The synthesized temperature record of China is consistent with that of northern hemisphere by Moberg et al. (2005). The present study is to examine possible responses of eco-environment to climate change over the last 2,000 years. The widely accepted climate records mentioned above provide basic comparable data for the present study.

## 1.3 Biotransport and Biovectors

Migratory organisms, e.g. seabirds, turtles and fish, can move across ecosystem boundaries, and act as biovectors transferring some substances from low potential area to high potential places against their concentrations (Fig. 1.3). This exerts significant impacts on distributions and biogeochemical cycling of some elements. Seabirds move between aquatic and terrestrial ecosystems, playing a critical role in exchanges of mass and energy between them. Minor changes in the concentrations of oceanic elements may induce significant alterations in their levels in bird

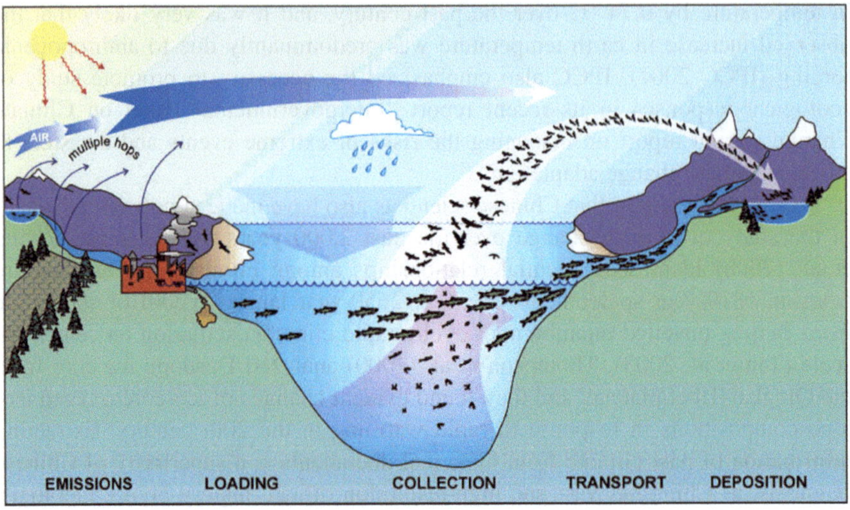


**Fig. 1.3** Schematic diagram of biological transport. [Reprinted with permission from Blais et al. (2007). Copyright (2007) American Chemistry Society]

tissues and droppings via food chains and biomagnification effects. This may further influence productivity, biodiversity and community structure of terrestrial ecosystems (Ellis 2005). The earlier studies of biotransport mainly concerned nutrient delivering between ecosystems. For instance, quantity of nutrients transferred from ocean to Heron Island of the Great Barrier Reef, Australia by seabirds was as high as 45 tons/year (Allaway and Ashford 1984). Such nutrients are extremely important for development of the coral island ecosystems. Scientists have found similar roles of turtles in Florida, and sea lions on the Galapagos Islands to that of seabirds on the Great Barrier Reef (Bouchard and Bjorndal 2000; Farina et al. 2003). Nutrients transported by biovectors are normally nitrogen and phosphorus. It should be mentioned that biogeochemical cycling of the element phosphorus is a non-gaseous process. Seabird is one of the a few ways that can transfer oceanic phosphorus to continents. Thus, seabird activity is of great significance to phosphorus cycling, due to their wide distributions and large population sizes. The Xisha Islands in the central South China Sea is famous habitats for numerous seabirds, whose activity has led to accumulation of large quantities of guano fertilizer on the islands. The Chinese government once performed several expeditions into the islands and exploited a great number of these guano fertilizers (Zhao 1996).

In recent years, scientists began to realize a severe problem associated with biotransport. Seabirds can not only transfer nutrients to the nutrient-poor island ecosystems, but also deliver significant quantities of contaminants, i.e., heavy metals (cadmium, mercury), persistent organic matters (POPs) etc., to pristine ecosystems. For example, Krummel et al. (2003) reported that levels of poly chlorinated biphenyl (PCB) were significantly enriched in sediments affected by salmon activity, almost a factor of seven higher than that of control areas, implying that migratory salmons transferred contaminants (such as POPs) from oceans to lacustrine sediments. In Arctic lakes, it has been found that lakes receive more pollutants, including HCB, DDT and mercury, where there are more seabirds, suggesting that seabirds can also greatly impact on concentrations of contaminants in high-latitude lacustrine ecosystems (Blais et al. 2005). Michelutti et al. (2009) also investigated compositions of Arctic lake sediments, and found that climate changes impacted on seabird population and further drove shift in Arctic lake ecosystems. Blais et al. (2007) reviewed various biovectors, and demonstrated that the effect of biovectors on long-rang transport of contaminants can exceed those of physical system (e.g., currents, winds), bringing about potential threat to the eco-environment. Despite of marked influences of seabirds on contaminants, efficacy of seabirds as metal biovectors may differ significantly, due to the differences in habitats, foraging behaviors, and dietary compositions. It has been reported that heavy metals, nutrient status, diatom taxa and elemental geochemistry characteristics of two Canadian ponds affected by Tern and Eider, respectively, are quite different. This suggests that seabirds may differ significantly in transport capacity, and biologically transported heavy metals may be species-specific (Michelutti et al. 2010).

## 1.4 Seabird Population Reconstruction

Population size is important for ecological study of seabirds, marine fish and mammalians. Following rationale of ecological geology, scientists reconstructed population records by animal remains in remote areas. For example, Sun et al. (2000) identified a group of avian bio-elements, including copper, zinc, cadmium, phosphorus, arsenic, calcium, strontium and fluorine, and reconstructed a 3000-year record of penguin population. From geochemical analysis, Liu et al. (2005) reconstructed a high-resolution 1300-year record of penguin population at Ardley Island, Antarctic, and identified that rapid decrease in numbers seemed to be caused by recent climate warming. Huang et al. (2009) reconstructed a longer record of penguin population extending back to 8500 years and investigated the "Holocene penguin optimum" across the whole Antarctica. By counting the number of subfossil penguin bones and radiocarbon analysis, Emslie et al. (2007) performed preliminary study of the occupation history of Adélie Penguin over the past 45,000 years, and also examined possible influences of climate change, sea-ice cover on abundance of this species. Liu et al. (2006a) reconstructed seabird population record at low-latitude area and hypothesized that a cool climate seemed more favorable to seabird on a tropical sea island. The geochemical approach to reconstructing population size has been extended from seabirds to mammalians and fish. For instance, a historical seal population at King George Island, Antarctica, over the last 1,500 years was estimated by Sun et al. (2004a). The results suggested that seal population tended to be linked to sea-ice coverage and temperature. Anthropogenic forcing is also a significant factor driving ecological changes. Yang et al. (2010)

emphasized remarkable influences of human activity, e.g. sealing industry, on dramatic decrease in seal population over the 20th century. Lacustrine sediments associated with migratory fish are enriched in $\delta^{15}N$ ($^{15}N/^{14}N$ relative to the standard) and $\delta^{15}N$ is an ideal marker for fish abundance, based on which Finney et al. (2002) reconstructed fisheries productivity in northern Pacific over the past 2,200 years. Liu et al. (2006b) analyzed stable isotope compositions of ornithogenic sediments, and also found that $\delta^{15}N$ was enriched in such sediments and thus could be used to track seabird activity. In terms of methodology, Liu et al. (2011) developed a new technique to reconstruct seabird population by spectral analysis of ornithogenic sediments. The spectrum-based record was similar to the bioelement-based record, testifying to its reliability. This enables a potential use of near-infrared reflectance spectroscopy (NIRS) to reconstruct seabird abundance rapidly.

Although the mechanism of long-term (centennial to millennial scale) changes in seabird ecology has not been well-documented, it is clear that seabird population can be affected by climate change. Scientists have made preliminary and tentative studies of the mechanism. These studies attempt to find the possible reasons by examining habitats, sea-ice coverage, food availability, human activity etc. A recent study by Cury et al. (2011) demonstrated global seabird responses to forage fish depletion, and hypothesized a minimal forage fish biomass to sustain seabird productivity. Via stable isotopic analysis, Huang et al. (2011) suggested that krill availability in the Southern Ocean inferred from fur seal $\delta^{15}N$ has been decreasing over the past century, and this was closely linked with decrease in sea ice extent. El Niño-Southern Oscillation (ENSO) may also be a factor that significantly affects seabirds (Fig. 1.4). For instance, Vandenbosch (2000) reported that the number of *Fregata minor* and *Sula sula* in Hawaii during the strong ENSO year 1982–1983 declined to 4 and 35%,

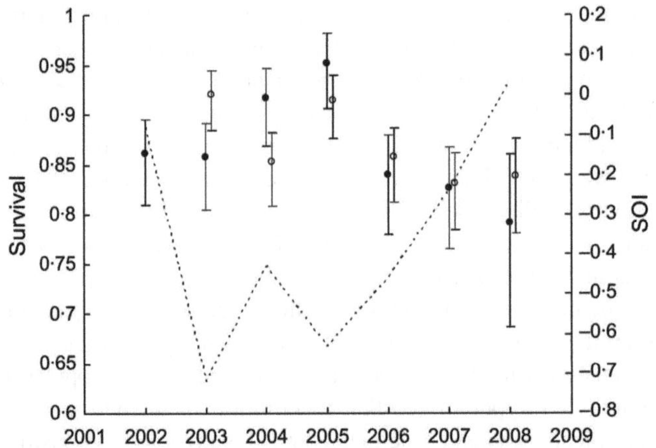


**Fig. 1.4** Shearwater survival in Pantaleu (*open circles*) and Chafarinas Islands (*filled circles*) since 2002 to 2008. Mean Southern Oscillation Index (SOI) is given as dotted line. Reprinted with permission from Contrasting effects of climatic variability on the demography of a trams-equatorial migratory seabird/Genovart M., Sanz-Aguilar A., Fernández-Chacón A., Igual J.M., Pradel R., Forero M.G., Oro D/Journal of Animal Ecology 82. Copyright [2013] [British Ecological Society]

respectively, of the levels before ENSO. The decrease in seabird population during ENSO years may be associated with marine productivity decline. In 1998, the ENSO was stronger than in normal years, leading to weaker East Asian monsoon and significantly lower surface primary productivity of the South China Sea (Zhao and Tang 2007). Additionally, over-fishing and seabird hunting may be also responsible for seabird population decrease. To discern different roles of climate change and human activity in determining seabird abundance is one of our objectives.

## 1.5 Palaeo-Diet Reconstruction

Reconstruction of palaeo-diet is of great importance in palaeoecology study. As reported in publications, there have been three approaches to reconstructing palaeo-diets. They are (1) morphological identification by prey remains, (2) DNA barcoding, and (3) stable isotope analysis. Of these three methods, morphological identification and DNA barcoding can reconstruct the exact dietary compositions of an organism, and work well for modern samples. Nonetheless, they are sometimes not applicable for ancient biological remains due to poor preservation and severe DNA degradation. In contrast, stable isotope analysis is a non-invasive technique and has been widely and successfully used to investigate diets of an organism.

### *1.5.1 Morphological Identification by Prey Remains*

The idea is to observe morphology of prey remains under low-power stereomicroscope, and then prey species and their relative abundances in the diets can be identified. For example, from observations of fish vertebra, otolith and squid beak in prey remains, it can be drawn the conclusion that diets of the predator include at least fish and squid. From this point, researchers investigated dietary compositions of Adélie penguins in the southern Ross Sea using prey remains recovered from ornithogenic sediments. 12 taxa of fish and two of squid with the Antarctic silverfish (*Pleuragramma antarcticum*) were identified as the most abundant prey species (Polito et al. 2002). A similar approach was applied to dietary reconstruction of penguins in East Antarctica over the past 9,000 years (Emslie and Woehler 2005). By morphological analysis, Lorenzini et al. (2009) inferred that the Adélie penguins selected prey averaging 67.23 ± 23 mm as their principal food during the Holocene (~10 ka).

### *1.5.2 DNA Barcoding*

DNA barcoding is a rapid, accurate and cost-effective means for species identification by genetic analysis. It involves the use of short and standard DNA segments to create sequences of known species against those of unknown (Hebert et al.

2003). The concept of DNA barcoding and most of representative work are from Professor Paul D.N. Hebert's research group in the University of Guelph, Canada. A standard DNA segment (648 base pairs, Cytochrome c oxidase subunit I, COI) in mitochondrial DNA is widely accepted and typically used as a DNA barcode for animals. However, it does not work well for plant species identifications. For plants, ribulose-bisphosphate carboxylase (*rbcL*) and maturase K (*matK*) in chloroplast DNA are considered as standard barcodes (Hebert et al. 2003; Jinbo et al. 2011). DNA barcoding has been widely used for bird species identifications, e.g., 260 species of North American birds and >200 species of Korean birds (Hebert et al. 2004; Yoo et al. 2006), as well as fish species identifications (Ward et al. 2005; Ivanova et al. 2007).

Seabird faeces is a special material, as it contains DNA from not only diets (e.g. fish, squid), but also host organism. We will also analyze the feasibility of isolating ancient DNA from subfossil guano particles in the present study.

### *1.5.3 Stable Isotope Analysis*

Stable isotope compositions, especially for light elements H, C, N, O etc., in different materials may differ significantly, due to isotope fractionation. Thus, they have been widely used for tracing, C3/C4-plant development examination, palaeodiet reconstruction and foraging behavior study (Zheng and Chen 2000). Of these light elements, stable carbon and nitrogen isotopes are ideal markers for foraging habitat and trophic position of an animal (Post 2002). Chamberlain et al. (2005) examined Pleistocene to recent dietary shifts in California Condor by stable carbon and nitrogen isotope analysis of bone collagen samples. A 9000-year isotope-based record of penguin diets was reconstructed by Emslie and Patterson (2007), and they found abrupt shift in penguin dietary compositions in recent 200 years. Norris et al. (2007) reported that estimated proportion of fish in murrelet diets was linked closely with murrelet abundance and trophic level over the past 40 years. The estimated proportion of krill in penguin diet by stable nitrogen isotope of seal hairs revealed its stepwise decrease over the past 100 years (Huang et al. 2011). Publication of Brian Fry's book "Stable Isotope Ecology" in 2007 is a milestone for the development of this discipline (Lin 2010). Another book namely "Tracking Animal Migration with Stable Isotope" written by Hobson and Wasenaar (Hobson and Wassenaar 2008) provided detailed descriptions of some typical stable isotopes, e.g. C, H, N, O, S, further improving its rationale and applications. Stable isotope ecology has drawn increasing attention in recent years. According to Newsome et al. (2010), both publication rate and total number of publications utilizing stable isotopes to study marine mammal ecology had substantially grown. However, only very few Chinese scientists are engaged in studies of stable isotope ecology.

Bird/bat guano is also an ideal material for the study of stable isotope ecology. Bats share similar morphology and foraging strategy with birds, and thus study of bat guano provides good reference for that of birds. Dr. Christopher M. Wurster from the University of St Andrews, Scotland and his colleagues consumed much time to study cave-dwelling bat with stable isotope analysis. They excavated bat guano deposits from a variety of places. They found that guano $\delta^{13}C$ could be used to track relative proportions of C3, C4 plants in bat diets, and further spatial vegetation distribution in the past. Taking guano samples from Southern America as an example, they identified an unambiguous rainfall-driven C3/C4 plant distribution (Wurster et al. 2007). Climate change throughout the Pleistocene-Holocene transition from $\delta^{13}C$ and $\delta D$ ($^{2}H/^{1}H$) values of guano in Bat Cave, Arizona was reconstructed, revealing cold Younger Dryas and 8.2 ka events (Wurster et al. 2008). In Southeast Asia, guano $\delta^{13}C$ analysis suggested significant change in the ratio of grass-derived carbon in bat diet, and remarkable forest contraction over the Last Glacial Period (Wurster et al. 2010). Following Wurster et al. (2010), Ayliffe (2010) hypothesized potential responses of guano isotope composition to millennial-scale climate variations in middle and high latitude areas.

## 1.6 The Study of Paleoceanography in the South China Sea

The South China Sea is a marginal sea, and quite sensitive to East Asian Monsoons. Seasonal changes in monsoon and ocean circulation render it an ideal place for studying alteration in marine environment. Marine sediments are by far the most widely used materials for studying paleoceanography. Using deep-sea deposits retrieved from the South China Sea, Wang et al. (1995) and Wang (1999) systematically studied ocean circulation, carbon cycling and climate change during late-Quaternary. Hu et al. (2000) reviewed ocean currents in the South China Sea, and emphasized the controlling role of East Asian Monsoon in surface circulation. Comprehensive study of paleoceanography in China began with the implementation of Ocean Drilling Program (ODP) in the South China Sea. Professor Wang Pinxian from the Tongji University, China co-led this international effort to conduct the first drilling expedition in the South China Sea in 1999. A total of 19 boreholes were drilled on continental slopes of the South China Sea to explore evolution the East Asian Monsoon (Qiu 2011). Following this international drilling program, Wang et al. (2003) reconstructed climate dynamics of the South China Sea back to as long as 32 million years. They further discussed the evolution and variability of the Asian monsoon system (Wang et al. 2005). Wei et al. (2007) reconstructed a record of SST over the last 260-ka years from Mg/Ca ratios of beneath foraminifer in the northern South China Sea. Marine sediments are normally low-resolution, and only few sediment cores from the South China Sea are on the millennial scale. Using high-resolution sediment cores from southern and northern slope of the South China Sea, Bian and Jian (2005) reconstructed SST and paleoproductivity records over the past 2,400 years. The results suggested a stepwise increase of marine productivity in the northern South China Sea, but gradual decrease in the southern part over the past 900 years.

High-resolution corals can be well absolute-dated, providing information about changes in marine environment on seasonal to annual scales. Researchers have used coral samples to reconstruct historical maritime records by geochemical analyses and U-Th dating, and have made advances in reconstructions of SST, precipitation, paleo-pH, sea level changes, East Asian Monsoon, etc. For example, Yu et al. (2002) analyzed of coral reef at Leizhou Peninsular, South China Sea, and revealed high-frequency, large-amplitude climatic oscillations during the Holocene. Sea level changes during middle to late Holocene were reconstructed by beach sediments and microatoll samples from northern South China Sea (Yu and Chen 2009; Yu et al. 2009). $^{14}C$ depleted material in oceans normally has a radiocarbon age older than their genuine age, known as reservoir effect. This is a troublesome problem in radiocarbon dating. Yu et al. (2010) estimated the reservoir effect in the South China Sea within the past 7,500 years and found significant variations in radiocarbon marine reservoir age. Sun et al. (2004b) hypothesized that Strontium in Porites coral from the Xisha Islands was an ideal proxy for SST. The reconstructed SST during the period from 1906 to 1994 showed that the late 20th century was approximately 1 °C warmer than the early 20th century. Stable oxygen isotope in corals has been identified to be linked closely with strength of East Asian Winter Monsoon, based on which Peng et al. (2003) reconstructed the record of winter monsoon over the past century. From boron isotope analysis of coral samples collected from Leizhou Peninsular, Hainan Island, Xisha and Nansha islands, Liu et al. (2009) found significant change in seawater paleo-pH of the South China Sea during the mid-late Holocene. Corals and historical documents also provide critical information about abrupt environmental events, including rapid and high-frequency cooling, typhoon landfalls and storms (Liu et al. 2001; Yu et al. 2002). More over, Shi et al. (2007) investigated previous studies on sea level changes in Southeast Asia, and documented the characteristics of sea level change associated with the South China Sea during the Holocene. These studies are of great significance for examining temperature changes of coastal China, flood/drought, and teleconnections among climatic factors.

In terms of resolution, marine sediments are generally low-resolution and suitable for the study of long-term environmental changes. Although coral has a high-resolution, it is quite difficult to find sequential coral samples, and thus more appropriate for short-term analysis. Our samples collect from the Xisha Islands fill the time interval between marine deposits and corals.

## 1.7 The Ecological Study of Coral Islands in the South China Sea

All the islands but one (Shidao Island), of the Xisha archipelago are coral islands derived from coral debris. Corals live in tropical oceans (Fig. 1.5) and are quite sensitive to ambient conditions, e.g. temperature and salinity. Climate change and human activity can significantly impact coral reefs, and thus they have been

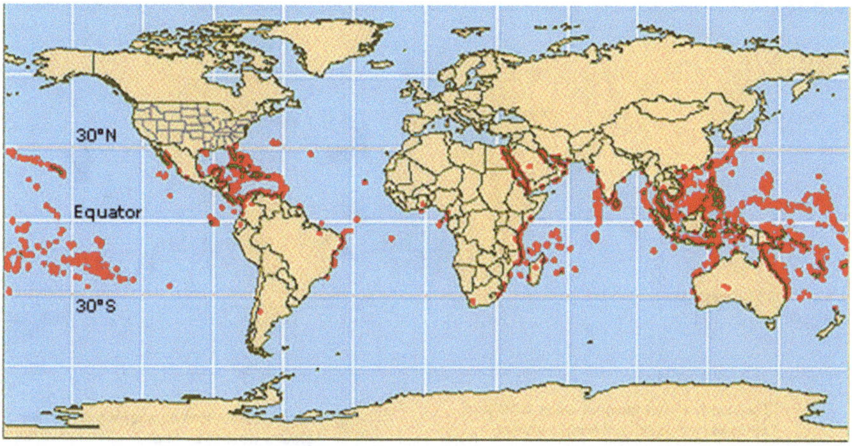


**Fig. 1.5** Worldwide distribution of coral reefs. Courtesy of the National Oceanic and Atmospheric Administration (http://oceanservice.noaa.gov/education/kits/corals/media/supp_coral05a.html)

increasingly used to study changes in tropical oceans in response to natural and anthropogenic forcing (Hughes et al. 2003). The Australian government initiated systematic investigation into the Great Barrier Reef and assessed its vulnerability and resilience under the context of global warming. In 2007, Great Barrier Reef Marine Park Authority and Department of the Environment and Heritage Australian Greenhouse Office released the synthesized report "Climate Change and the Great Barrier Reef: A vulnerability Assessment", in which scientists emphasize significant influences of climate change on various aspects of coral reef ecosystems. For instance, warming has resulted in coral bleaching, plant destruction, mammalian breeding behavior change and seabird population decrease (Hoegh-Guldberg et al. 2007; Turner and Batianoff 2007; Hamann et al. 2007; Congdon et al. 2007).

### *1.7.1 The Great Barrier Reef*

The Great Barrier Reef, a part of coastal Australia, is the largest coral reef system in the world. It is composed of more than 2900 individual reefs and covers >340,000 km$^2$ (Wachenfeld et al. 2007). Though the Great Barrier Reef is far away from the South China Sea, the ecological study of it provides good reference for the Xisha Islands, due to their high similarity in geological settings, as well as biodiversity. It is likely that climate change impacts on marine environment and further causes changes in population size, reproductive behavior and foraging strategy of higher organisms, e.g. birds, amphibians and marine mammals (Fig. 1.6). A weakness of the ecological studies of the Great Barrier Reef is that scientists focus mainly on modern observations, so a long-term ecological record, and comparisons between the present and the past are extremely insufficient.

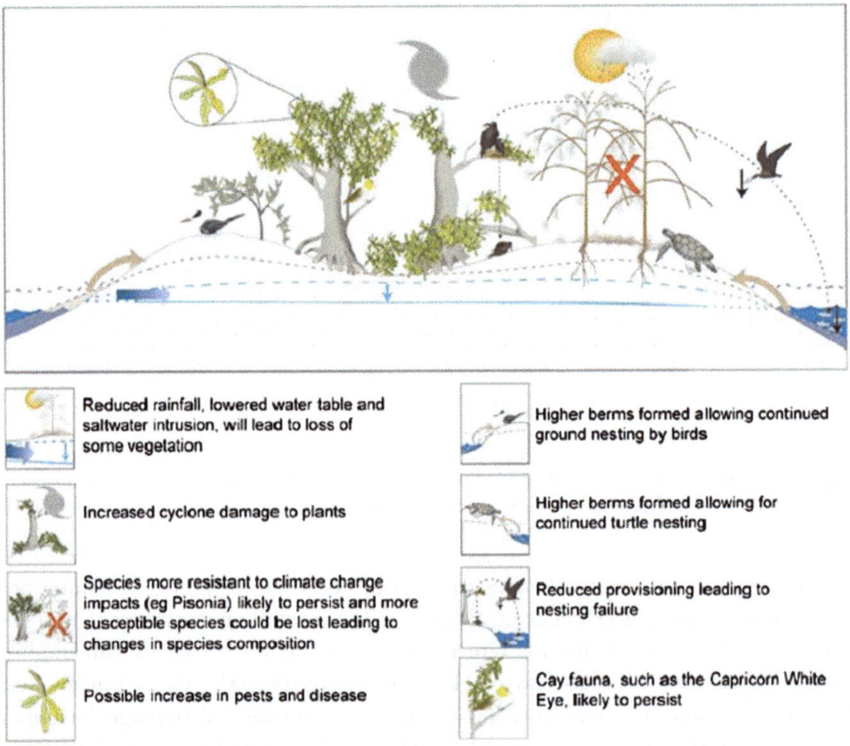


**Fig. 1.6** Potential impacts of climate change on coral reef ecosystems (Turner and Batianoff 2007, reproduced with permission from Great Barrier Reef Marine Park Authority)

### *1.7.2 History of Expeditions to the Xisha Islands*

Chinese ancestors discovered the Xisha Islands in as early as the Late Neolitic Age, and began to settle in this area at least 2,000 years ago (Zhao 1996). However, scientists did not perform genuine scientific survey of the Xisha archipelago until the 1920–30s. Early studies of the Xisha Islands area mainly concerned with environmental background data. With advances in science and technology, research topics have changed gradually from individual subjects to interdisciplinary ecological geology. Investigation groups involve government agencies, People's Liberation Army, academic institutions and universities. A brief summary of the expeditions to the Xisha Islands is supplied in Table 1.1.

Boreholes Xichen-1, Xiyong-1 and Xishi-1, were collected from Chenhang Island, Yongxing Island and Shidao Island, respectively, in 1983–84. Formation and micropaleontology of such islands were then investigated in detail, promoting the study of coral reef deposits (Zhang et al. 1989). Scientists drilled similar boreholes in Nansha Islands and examined correlations between sedimentation rate of coral debris and climate change (Zhao et al. 2000). A 1670-year record of

**Table 1.1** Brief summary of scientific expeditions to the Xisha Islands[a]

| Year | Organizer | Content | Main achievement |
|---|---|---|---|
| 1928 | The Xisha Islands Investigation Committee | Geography, culture, current, climate, resource etc. | Book: Investigation report of the Xisha Islands |
| 1947 | Academia Sinica, Sun Yat-Sen University | Geology, phosphorite, vegetation, soil etc. | Research articles |
| 1973–1974 | South China Sea Institute of Oceanology, CAS | Biology, geophysics, hydrology etc. | Book: Biological resources in the Xisha and Zhongsha islands |
| 1974 | Institute of Soil Science CAS; Institute of Zoology CAS; South China Institute of Botany; Institute of Geography CAS; Xiamen University | Soil, phosphorit, biology, sovereignty etc. | Book: Soil and Guano Phosphorus in Xi-Sha.<br>Plants and vegetation on Xisha Islands of China<br>Research Articles |
| 1978 | South China Sea Institute of Oceanology, CAS; South China Sea Fleet | Geology, geomorphology, etc. | Unpublished |
| 1983–1984 | Institute of Marine Geology, Ministry of Geology and Mineral Resources | Microfossil, geology, minerals etc. | Book: Study of sedimentary geology of coral reef in Xisha |
| 1991 | Hainan Ocean Administration | Climate, soil, biology, resource etc. | Book: The professional proceedings of the integrated investigation research on sea islands resource of Hainan province |
| 2003–2004, 2008 | University of Science and Technology of China | Seabirds, mammals, vegetation, geology etc. | Research articles |

[a] Compiled from Zhao 1996, 2006; Exploration Group of Xisha Islands of Institute of Soil Science CAS Exploration Group of Xisha Islands of Institute of Soil Science of Chinese Academy of Sciences (CAS) 1977; Hainan Ocean Administration 1999; Zhang et al. 1989

temperature in the Nansha Islands was also reconstructed by stable oxygen isotope ($\delta^{18}O$) of foraminifer in these sediments (Zhao et al. 2004). The results showed that temperature in the South China Sea was a response to climate change in China and the world. From analysis of Sr/Ca and Mg/Ca of corals, two high-resolution short SST records during the Holocene were reconstructed, suggesting a decrease of SST by approximately 1 °C 540 years ago (the Little Ice Age) (Wei et al. 2004). Liu et al. (2008c) studied time-series of SST in the coastal Xisha Islands and Hainan Island and reported a Kuroshio-driven SST change, possibly a result of decline in the strength of East Asian Winter Monsoon over the past 100 years. We have performed two investigations into the ecological geology of the Xisha Islands in 2003 and 2008, respectively. These field studies facilitated our understanding of historical seabird population, environmental quality (Xie et al. 2005), greenhouse gases emission (Zhu et al. 2005) etc., in the Xisha Islands. For example, Liu et al. (2006a; 2008a, b) reconstructed a 1300-year record of seabird population, as well as a 1100-year record of precipitation, and examined paleoenvironmental implications of the guano phosphatic cementation on Dongdao Island, by geochemical analysis and radiocarbon dating. Yan et al. (2011a, b) identified that mean grain size of lacustrine sediments on Dongdao Island is an ideal marker for historical precipitation, and hypothesized a new approach for tracking palaeo-ENSO. The present study is conducted on the basis of these studies, and aims to decode mechanism of seabird ecology changes in response to climate change and human activity.

### *1.7.3 Modern Observations of Seabird Ecology on the Xisha Islands*

The Xisha archipelago is well-known for its numerous seabirds. Only few researchers have investigated the modern seabird ecology, e.g. population size, foraging strategy etc., in the Xisha archipelago. Except for our group, none of previous scientists have focused on palaeoecology. Modern observations of these birds are of prime importance for examining seabird ecology in the past. According to demographics, a variety of seabird species, including Red-footed Booby, Brown Booby, Sooty Tern etc., inhabited the Xisha Islands in large numbers (Cao et al. 2007). Bei and Tang (1959) reported that Red-footed Booby disappeared on Yongxing Island due to hunting and egg collection. After that, Institute of Zoology, CAS, investigated biodiversity of the Xisha Island in 1960–70s. According to statistics, there were more than 60 species of birds on these islands in 1970s (Exploration Group of Xisha Islands of Institute of Soil Science of Chinese Academy of Sciences 1977). Since 2003, Dr. Cao Lei's research group from University of Science and Technology of China, have investigated population size, foraging behavior, dietary composition and reproductive strategy of seabirds in the Xisha Islands for several years. They reported a significant decrease in seabird population over the last decades (Cao 2005; Cao et al. 2005, 2007). The reasons, however, remain unknown and needs further in-depth discussion.

The above-mentioned studies of paleoclimatology, paleoceanography and modern seabird ecology of the Xisha Islands provide important background information and comparison data for the present Ph.D. study. At the same time, the ecological response of the Xisha Islands to climate and human activity are poorly understood, and a further detailed discussion is imperative.

## References

Allaway WG, Ashford AE (1984) Nutrient input by seabirds to the forest on a coral island of the Great Barrier Reef. Mar Ecol Prog Ser 19:297–298

Ayliffe LK (2010) How to sample the carbon isotopes of tropical ecosystems without leaving your armchair. Proc Natl Sci USA 107:15664–15665

Bei TX, Tang ZM (1959) Observation on the fertilizer resource of Yongxing Island, Xisha Archipelago. Chin J Zool 3:317–319

Bian YP, Jian Z (2005) Paleoceanographic changes in the South China Sea over the last 2400 years and their comparison with the historical paleoclimatical records. Mar Geol Quat Geol 25(4):73–78

Blais JM, Kimpe LE, McMahon D, Keatley BE, Mallory ML, Douglas MSV (2005) Arctic seabirds transport marine-derived contaminants. Science 309:445

Blais JM, Macdonald RW, Mackay D, Mackay D, Webster E, Harvey C, Smol JP (2007) Biologically mediated transport of contaminants to aquatic systems. Environ Sci Technol 41:1075–1084

Bouchard SS, Bjorndal KA (2000) Sea turtles as biological transporters of nutrients and energy from marine to terrestrial ecosystems. Ecology 81(8):2305–2313

Cao L (2005) Population ecology of the red-footed booby on the Xisha Archipelago. Ph.D. dissertation, Lanzhou University

Cao L, Pan YL, Liu NF (2005) Status of the red-footed booby on the Xisha Archipelago, South China Sea. Waterbirds 28(4):411–419

Cao L, Pan YL, Liu NF (2007) Waterbirds of the Xisha Archipelago, South China Sea. Waterbirds 30(2):296–300

Chamberlain CP, Waldbauer JR, Fox-Dobbs K, Newsome SD, Koch PL, Smith DR, Chruch ME, Chamberlain SD, Sorenson KJ, Risebrough R (2005) Pleistocene to recent dietary shifts in California condors. Proc Natl Acad Sci USA 102:16707–16711

Chu CZ (1972) A preliminary study on the climatic fluctuations during the last 5000 years in China. Acta Archaeologica Sinica 1:15–38

Congdon BC, Erwin CA, Peck DR et al (2007) Vulnerability of seabirds on the Great Barrier Reef to climate change. In: Marshall PA, Johnson J (eds) Climate change and the Great Barrier Reef. Great Barrier Reef Marine Park Authority, Townsville, Australia

Cury PM, Boyd IL, Bonhommeau S, Anker-Nilssen T, Crawford RJM, Furness RW, Mills JA, Murphy EJ, Österblom H, Paleczny M, Piatt JF, Roux JP, Shannon L, Sydeman WJ (2011) Global seabird response to forage fish depletion—one-third for the birds. Science 234:1703–1706

Ellis JC (2005) Marine birds on land: a review of plant biomass, species richness, and community composition in seabird colonies. Plant Ecol 181:227–241

Emslie SD, Patterson WP (2007) Abrupt recent shift in $\delta^{13}C$ and $\delta^{15}N$ values in Adelie penguin eggshell in Antarctica. Proc Natl Acad Sci 104:10213–10217

Emslie SD, Woehler EJ (2005) A 9000-year record of Adélie penguin occupation and diet in the Windmill Islands, East Antarctica. Antarctic Science 17(1):57–66

Emslie SD, Coats L, Licht K (2007) A 45,000 year record of Adélie Penguins and climate change in the Ross Sea, Antarctica. Geology 35:61–64

Exploration Group of Xisha Islands of Institute of Soil Science of Chinese Academy of Sciences (CAS) (1977) Soil and guano phosphorus mine in Xi-Sha. Science Press, Beijing

Farina JM, Salazar S, Wallem KP, Witman JD, Ellis JC (2003) Nutrient exchanges between marine and terrestrial ecosystems: the case of the Galapagos sea lion Zalophus wollebaecki. J Anim Ecol 72:873–887

Figueirido B, Janis CM, Perez-Claros JA, Renzi MD, Palmqvist P (2012) Cenozoic climate change influences mammalian evolutionary dynamics. Proc Natl Acad Sci USA 109(3):722–727

Finney BP, Gregory-Eaves I, Douglas MSV, Smol JP (2002) Fisheries productivity in the northeastern Pacific Ocean over the past 2,200 years. Nature 416:729–733

Hainan Ocean Administration (1999) The professional proceedings of the integrated investigation research on sea islands resource of Hainan province. China Ocean Press, Beijing

Hamann M, Limpus CJ, Read MA (2007) Vulnerability of marine reptiles on the Great Barrier Reef to climate change. In: Marshall PA, Johnson J (eds) Climate change and the Great Barrier Reef. Great Barrier Reef Marine Park Authority, Townsville, Australia

Hebert PDN, Cywinska A, Ball SL, deWarrd JR (2003) Biological identifications through DNA barcodes. Proc R Soc Lond B 270:313–321

Hebert PDN, Stoeckle MY, Zemlak TS, Francis CM (2004) Identification of birds through DNA barcodes. PLoS Biol 2:e312

Hobson KA, Wassenaar LI (2008) Tracking animal migration with stable isotopes. Academic Press, California

Hoegh-Guldberg O, Anthony K, Berkelmans R, Dove S, Fabricus K, Lough J, Marshall P, Oppen M, Negri A, Willis B (2007) Vulnerability of reef-building corals on the Great Barrier Reef to climate change. In: Marshall PA, Johnson J (eds) Climate change and the Great Barrier Reef. Great Barrier Reef Marine Park Authority, Townsville, Australia

Hu JH, Kawamura H, Hong HS, Qi YQ (2000) A review on the currents China Sea: seasonal circulation, South China Sea warm current and Kuroshio intrusion. J Oceanogr 56:607–624

Hu FS, Brown TA, Curry BB, Engstrom DR (2001) Pronounced climatic variations in Alaska during the last two millennia. Proc Natl Acad Sci USA 98(19):10552–10556

Huang T, Sun LG, Wang YH, Liu XD, Zhu RB (2009) Penguin population dynamics for the past 8500 years at Gardner Island, Vestfold Hills. Antarct Sci 21(6):571–578

Huang T, Sun LG, Stark J, Wang Y, Cheng ZQ, Yang QC, Sun S (2011) Relative changes in krill abundance from Antarctic fur seal. PLoS ONE 6(11):e27331. doi:10.1371/journal.pone.0027331

Hughes TP, Baird AH, Bellwood DR, Card M, Connolly SR, Folke C, Grosberg R, Hoegh-Guldberg O, Jackson JBC, Kleypas J, Lough JM, Marshall P, Nystrom M, Palumbi SR, Pandolfi JM, Rosen B, Roughgarden J (2003) Climate change, human impacts, and the resilience of coral reefs. Science 301(5635):929–933

IPCC (2007) Climate change 2007: synthesis report

Ivanova NV, Zemlak TS, Hanner RH, Hebert PDN (2007) Universal primer cocktails for fish DNA barcoding. Mol Ecol Notes 7(4):544–548

Jinbo U, Kato T, Ito M (2011) Current progress in DNA barcoding and future implications for entomology. Entomol Sci 14:107–124

Jones PD, Mann ME (2004) Climate over past millennia. Rev Geophys 42:RG2002

Krummel EM, Macdonald RW, Kimpe LE, Gregory-Eaves I, Demers MJ, Smol JP, Finney B, Blais JM (2003) Delivery of pollutants by spawning salmon. Nature 425:255–256

Lambert DM, Ritchie PA, Millar CD, Holland B, Drummond AJ, Baroni C (2002) Rates of evolution in ancient DNA from Adelie Penguins. Science 295:2270–2273

Lin GH (2010) Stable isotope ecology: a new branch of ecology resulted from technology advances. Chin J Plant Ecol 34(2):119–122

Liu TS (1997) Quaternary environment. Science Press, Beijing

Liu KB, Shen CM, Louie KS (2001) A 1,100-year history of typhoon landfalls in Guangdong, Southern China, reconstructed from Chinese historical documentary records. Ann Assoc Am Geogr 91(3):453–464

Liu XD, Sun LG, Xie ZQ, Yin XB, Wang YH (2005) A 1300-year record of penguin populations at Ardley Island in the Antarctic, as deduced from the geochemical data in the ornithogenic lake sediments. Arct Antarct Alp Res 37(4):490–498

Liu XD, Zhao SP, Sun LG, Luo HH, Yin XB, Xie ZQ, Wang YH, Liu KX, Hu XH, Ding XF, Fu DP (2006a) Geochemical evidence for the variation of historical seabird population on Dongdao Island of the South China Sea. J Paleolimnol 36:259–279

Liu XD, Li HC, Sun LG, Yin XB, Zhao SP, Wang YH (2006b) $\delta^{13}C$ and $\delta^{15}N$ in the Ornithogenic sediments from the Antarctic Maritime as Paleoecological Proxies during the past 2000 Years. Earth Planet Sci Lett 243:424–438

Liu XD, Sun LG, Wei GJ, Wang YH, Yan H, Liu KX, Wu XH (2008a) A 1100-year palaeoenvironmental record infered from stable isotope and trace element compositions of ostracode and plant caryopses in sediments of Cattle Pond, Dongdao Island, South China Sea. J Paleolimnol 2008(40):987–1002

Liu XD, Zhao SP, Sun LG, Yin XB, Xie ZQ, Wang YH (2008b) Paleoenvironmental implications of the Guano phosphatic cementation of coral calcarenite on the Dongdao Island in the South China Sea. Mar Geol 247:1–16

Liu Y, Peng ZC, Chen TG, Wei GJ, Sun WD, Sun RY, He JF, Liu GJ, Chou CL, Zartman RE (2008c) The decline of winter monsoon velocity in the South China Sea through the 20th century: evidence from the Sr/Ca records in corals. Glob Planet Change 63:79–85

Liu Y, Liu WG, Peng ZC, Xiao YK, Wei GJ, Sun WD, He JF, Liu GJ, Chou CL (2009) Instability of seawater pH in the South China Sea during the mid-late Holocene: Evidence from boron isotopic composition of corals. Geochim Cosmochim Acta 73:1264–1272

Liu XD, Sun J, Sun LG, Liu WQ, Wang YH (2011) Reflectance spectroscopy: a new approach for reconstructing penguin population size from Antarctic ornithogenic sediments. J Paleolimnol 45:213–222

Lorenzini S, Olmastroni S, Pezzo F, Salvatore MC, Baroni C (2009) Holocene Adelie penguin diet in Victoria Land, Antarctica. Polar Biol 32:1077–1086

Mann EE, Jones PD (2003) Global surface temperatures over the past two millennia. Geophys Res Lett 30(15):5-1–5-4

Mann ME, Bradley RS, Hughes MK (1999) Northern Hemisphere temperatures during the past millennium: Inferences, Uncertainties, and Limitations. Geophys Res Lett 26(6):759–762

Mann ME, Zhang ZH, Kughes MK, Bradley RS, Miller SK, Rutherford S, Ni FB (2008) Proxy-based reconstructions of hemispheric and global surface temperature variations over the past two millennia. Proc Natl Acad Sci USA 105:13252–13257

Michelutti N, Keatley BE, Brimble S, Blais JM, Liu HJ, Douglas SV, Mallory ML, Macdonald RW, Smol JP (2009) Seabird-driven shifts in Arctic pond ecosystems. Proc R Soc B 276:591–596

Michelutti N, Blais JM, Mallory ML, Brash J, Thienpont J, Kimpe LE, Douglas MSV, Smol JP (2010) Trophic position influences the efficacy of seabirds as metal biovectors. Proc Natl Acad Sci USA 107:10543–10548

Moberg A, Sonechldn Dmitry M, Karin Holmgren, Datsenko Nina M, Wibjorn Karlen (2005) Highly variable Northern Hemisphere temperatures reconstructed from low- and high-resolution proxy data. Nature 433:613–617

National Research Council of the National Academies (2006) Surface temperature reconstructions for the last 2000 Years. The National Academies Press, Washington, D.C.

Newsome SD, Clementz MT, Koch PL (2010) Using stable isotope biogeochemistry to study marine mammal ecology. Mar Mammal Sci 26(3):509–572

Norris DR, Arcese P, Preikshot D, Bertram DF, Kyser TK (2007) Diet reconstruction and historic population dynamics in a threatened seabird. J Appl Ecol 44:875–884

Oppo DW, Rosenthal Y, Linsley BK (2009) 2,000-year-long temperature and hydrology reconstructions from the Indo-Pacific warm pool. Nature 460:1113–1116

Peng ZC, Chen TG, Nie BF, Head MJ, He XX, Zhou WJ (2003) Coral $\delta^{18}O$ records as an indicator of winter monsoon intensity in the South China Sea. Quatern Res 59(3):285–292

Polito M, Emslie SD, Walker W (2002) A 1000-year record of Adelie penguin diets in the southern Ross Sea. Antarct Sci 14(4):327–332

Post DM (2002) Using stable isotopes to estimate trophic position: models, methods and assumptions. Ecology 83:703–718

Qiu J (2011) China's unsinkable scientist. Nature 469:460–461

Shi XJ, Yu KF, Chen TG (2007) Progress in researches on sea-level changes in South China Sea since mid-Holocene. Mar Geol Quat Geol 27(5):121–132

Sicre M, Jacob J, Ezat U, Rousse S, Kissel C, Yiou P, Eiriksson J, Knudsen KL, Jansen E, Turon J (2008) Decadal variability of sea surface temperatures off North Iceland over the last 2000 years. Earth Planet Sci Lett 268:137–142

Sun LG, Xie ZQ, Zhao JL (2000) A 3000-year record of penguin population. Nature 407:808

Sun LG, Liu XD, Yin XB, Zhu RB, Xie ZQ, Wang YH (2004a) A 1500-year record of Antarctic seal populations in response to climate. Polar Biol 27:495–501

Sun YL, Sun M, Wei GJ, Lee T, Nie BF, Yu ZW (2004b) Strontium contents of a Porites coral from Xisha Island, South China Sea: a proxy for sea-surface temperature of the 20th century. Paleoceanography 19:PA2004. doi:10.1029/2003PA00959

Sun LG, Yin XB, Liu XD, Zhu RB, Xie ZQ, Wang YH (2006) A 2000-year record of mercury and ancient civilizations in seal hairs from King George Island, West Antarctica. Sci Total Environ 368:236–247

Tan M, Liu TS, Hou JZ, Qin XG, Zhang HC, Li TY (2003) Cyclic rapid warming on centennial-scale revealed by a 2650-year stalagmite record of warm season temperature. Geophys Res Lett 30(12):1617. doi:10.1029/2003GL017352

Thompson LG, Yao T, Mosley-Thompson E, Davis ME, Henderson KA, Lin PN (2000) A high-resolution millennial record of the south Asian monsoon from Himalayan ice cores. Science 289:1916–1919

Turner M, Batianoff GN (2007) Vulnerability of island flora and fauna in the Great Barrier Reef to climate change. In: Marshall PA, Johnson J (eds) Climate change and the Great Barrier Reef. Great Barrier Reef Marine Park Authority, Townsville, Australia

Tyson PD, Lindesay JA (1992) The climate of the last 2000 years in southern Africa. The Holocene 2(3):271–278

Vandenbosch R (2000) Effects of ENSO and PDO events on seabird population as revealed by Christmas bird count. Waterbirds 23(3):416–422

Wachenfeld D, Johnson J, Skeat A, Kenchington R, Marshall P, Innes J (2007) Introduction to the Great Barrier Reef and climate change. In: Marshall PA, Johnson J (eds) Great Barrier Reef ecological vulnerability assessment. Great Barrier Reef Marine Park Authority, Townsville, Australia

Wang PX (1999) Response of Western Pacific marginal seas to glacial cycles: paleoceanographic and sedimentological features. Mar Geol 156:5–39

Wang PX, Wang LJ, Bian YH, Jian ZM (1995) Late Quaternary paleoceanography of the South China Sea: surface circulation and carbonate cycle. Mar Geol 127:145–165

Wang PX, Zhao QH, Jian ZM, Cheng XR, Huang W, Tian J, Wang JL, Li QY, Li BH, Su X (2003) Third million year deep sea records in the South China Sea. Chin Sci Bull 48(23):2524–2535

Wang PX, Clemens S, Beaufort L, Braconnot P, Ganssen G, Jian ZM, Kershaw P, Sarnthein M (2005) Evolution and variability of the Asian monsoon system issues. Quatern Sci Rev 24:595–629

Ward RD, Zemlak TS, Innes BH, Last PR, Hebert PDN (2005) DNA barcoding Australia's fish species. Phil Trans R Soc B 360:1847–1857

Wei GJ, Yu KF, Li XH, Zhao JX, Sun YL, Sun M, Nie BF (2004) Coralline Sr/Ca and Mg/Ca thermometer for the northern South China Sea: calibration and primary application on high resolution SST reconstructing. Quat Sci 24(3):325–331

Wei GJ, Deng WF, Liu Y, Li XH (2007) High-resolution sea surface temperature records derived from foraminiferal Mg/Ca ratios during the last 260 ka in the northern South China Sea. Palaeogeogr Palaeoclimatol Palaeoecol 250:126–138

Wurster CM, McFarlane DA, Bird MI (2007) Spatial and temporal expression of vegetation and atmospheric variability from stable carbon and nitrogen isotope analysis of bat guano in the southern United States. Geochim Cosmochim Acta 71:3302–3310

Wurster CM, Patterson WP, McFarlane DA, Wassenaar LI, Hobson KA, Athfield NB, Bird MI (2008) Stable carbon and hydrogen isotopes from bat guano in the Grand Canyon, USA, reveal Younger Dryas and 8.2 ka events. Geology 36(9):683–688

Wurster CM, Bird MI, Bull LD, Creed F, Bryant C, Dungait JAJ (2010) Forest contraction in north equatorial Southeast Asia during the Last Glacial Period. Proc Natl Acad Sci USA 106:11188–11193

Xie ZQ, Sun LG, Zhang PF, Zhao SP, Yin XB, Liu XD, Cheng BB (2005) Preliminary geochemical evidence of groundwater contamination in coral islands of Xisha, South China Sea. Appl Geochem 20(10):1848–1856

Yan H, Sun LG, Oppo DW, Wang YH, Liu ZH, Xie ZQ, Liu XD, Cheng WH (2011a) South China Sea hydrological changes and Pacific Walker Circulation variations over the last millennium. Nat Commun 2:293. doi:10.1038/ncomms1297

Yan H, Sun LG, Wang YH, Huang W, Qiu SC, Yang CY (2011b) A record of the Southern Oscillation Index for the past 2,000 years from precipitation proxies. Nat Geosci. doi:10.1038/ngeo1231
Yang B, Braeuning A, Kathleen R, Johnson KR, Shi YF (2002) General characteristics of temperature variation in China during the last two millennia. Geophys Res Lett 29(9):38-1–38-4
Yang QC, Sun LG, Kong DM, Huang T, Wang YH (2010) Variation of Antarctic seal population in response to human activities in 20th century. Chin Sci Bull 55(11):1084–1087
Yoo HS, Eah JY, Kim JS, Kim YJ, Min MS, Paek WK, Lee H, Kim CB (2006) DNA barcoding Korean birds. Mol Cells 22:323–327
Yu KF, Chen T (2009) Beach sediments from northern South China Sea suggest high and oscillating sea levels during the late Holocene. Earth Sci Front 16(6):138–145
Yu KF, Liu DS, Shen CD, Zhao JX, Chen TG, Zhong JL, Zhao HT, Song CJ (2002) High-frequency climatic oscillations recorded in a Holocene coral reef at Leizhou Peninsula, South China Sea. Sci China (Ser D) 45(12):1057–1067
Yu KF, Zhao JX, Done T, Chen TG (2009) Microatoll record for large century-scale sea-level fluctuations in the mid-Holocene. Quatern Res 71:354–360
Yu KF, Hua Q, Zhao JX, Hodge E, Barbetti M (2010) Holocene marine $^{14}C$ reservoir age variability: evidence from 230Th-dated corals in the South China Sea. Paleoceanography 25:PA3205. doi:10.1029/2009PA001831
Zhang DE (2004) A Compendium of Chinese meterological records of the last 3000 years. Jiangsu Education Publishing House, Nanjing
Zhang DE, Lu LH (2007) Anti-correlation of summer/winter monsoon. Nature 450:E7–E8
Zhang MS, He QX, Ye ZZ, Han CR, Li H, Wu JZ, Ju LJ (1989) Study on sedimentary geology of biological reefs and carbonate in the Xisha Islands. Science Press, Beijing
Zhang PZ, Cheng H, Edwards RL, Chen FH, Wang YJ, Yang XL, Liu J, Tan M, Wang XF, Liu JH, An CL, Dai ZB, Zhou J, Zhang DZ, Jia JH, Jin LY, Johnson KR (2008) A test of climate, sun and culture relationships from an 1810-year Chinese cave record. Science 322:940–942
Zhao HT (1996) History of expeditions to Xisha Islands. Geogr Res 15:55–65
Zhao SP (2006) Evolution of seabird eco-environment on Xisha Islands of South China Sea. Master thesis. University of Science and Technology of China
Zhao H, Tang DL (2007) Effect of 1998 El Niño on the distribution of phytoplankton in the South China Sea. J Geophys Res 112:C0217
Zhao HT, Wen XS, Wang LR (2000) Climate change and the depositional velocity in the lagoon of Yongshu atoll, Nansha Islands. Trop Geogr 20(4):247–249
Zhao HT, Wen XS, Wang LR, Peng ZC (2004) The temperature changes recorded as $\delta^{18}O$ in the lagoon sediment of the Nansha Islands in the past 1670 years. Trop Geogr 24(2):103–108
Zheng YF, Chen JF (2000) Stable isotope geochemistry. Science Press, Beijing
Zhu RB, Sun LG, Zhao SP, Xie ZQ, Liu XD, Yin XB (2005) Preliminary studies on methane flux from the ornithogenic soils on Xi-sha atoll, South China Sea. J Environ Sci 17(5):783–793

# Chapter 2
# Research Contents and Methodology

## 2.1 Background and Research Significance

### *2.1.1 Background*

With the rapid development of the human society since the Industrial Revolution, more and more global environmental problems, e.g. global warming, sea level rise, ozone depletion, frequently extreme weathers, etc., occur to unprecedented amplitudes, for which we pay a painful price. Especially since the 1960s, global change has gradually become the limitation factor in social-economic development. To promote the study of global changes and broad cooperations among countries, International Human Dimensions Programme on Global Environmental Change (IHDP), International Geosphere Biosphere Programme (IGBP), World Climate Research Programme (WCRP) and DIVERSITAS jointed together to found the Earth System Science Partnership (ESSP) in 2001. Global warming is one of the most typical issues of global changes, and has drawn increasing attentions from politics, academia and ordinary citizens worldwide. According to the 4th synthesized report of climate change study released by Intergovernmental Panel on Climate Change (IPCC) in 2007, global surface temperature over the last century has increased by 0.74 °C. Continuous increase in temperature exerts significant impacts on many aspects of natural systems and human society (IPCC 2007). Changes of eco-environments in response to climate change and anthropogenic forcing has become focus and frontier of global change studies. To better understand ecological response to global changes on a scientific basis and better adapt to those changes become urgent. Study of "Human activities and their impacts on the Earth system" and "Global change and regional response" have also been listed as Basic Research in Response to Major National Strategic Needs in "The National Medium- and Long-Term Program for Science and Technology Development (2006–2020)" in China.



L. Xu, *Impact of Climate Change and Human Activity on the Eco-environment*,
Springer Theses, DOI 10.1007/978-3-662-45003-1_2

Understanding the past is of great importance for predicting the future. Scientists have used a variety of natural archives, e.g. polar/alpine glaciers, marine sediments, loess deposits, speleothems (stalagmite), lacustrine sediments, etc., to study palaeo-climate and palaeo-ecology on different time-scales (Petit et al. 1999; Thompson et al. 2000; Wang 1999; Guo et al. 2002; Wang et al. 2008; Liu and Herbert 2004). Those studies provide essential information for discriminating different roles of natural and human forcing, and predicting the future. However, differences in ecosystem-type, and complexity of climate systems render future prediction quite uncertain. For example, scientists found that seabirds (penguins) in high-latitude Antarctica tended to prefer warm climate (Sun et al. 2000). In contrast, a cool climate seemed more favorable to seabirds in tropical areas (Liu et al. 2006). Some key issues remain unclear. For instance, coupling between climate and productivity, and impacts of climate dynamics on ecosystems have not been well-documented and need further in-depth discussion. To better understand these processes, scientists initiated some International research programs, including International Geosphere Biosphere Programme (IGBP), World Climate Research Programme (WCRP) and WCRP-based Climate Variability and Predictability (CLIVAR).

In terms of time scale, palaeo-climate and palaeo-ecology over the past two millennia are of particular interest (Shi 1997), as it is not only highly related to human civilizations, but also susceptible to be affected by anthropogenic feedbacks. The study of climate during this time thus provides a "key" to predict the future, and has been emphasized by international organizations. Past Global Changes (PAGES) aims at understanding the Earth's past environment and making predictions for the future. The study of climate in the last two millennia is one of PAGES's two focuses. National Research Council of the Nation Academies, USA (2006) also released its synthesized report namely "Surface Temperature Reconstruction for the Last 2,000 years". The present study aims to reconstruct ecological records of the Xisha Islands, South China Sea, over the past 2,000 years. Via multi-proxy analysis and regional comparisons, it is attempted to decode possible interactions among climate change, human activity and ecosystems. This is helpful for the environmental conservation, resource exploitation and future climate prediction in the South China Sea.

### *2.1.2 Significance of the Present Work*

Seabirds frequently move across aquatic and terrestrial ecosystem boundaries, playing an important role in exchange of matter and energy between ecosystems. Those biovectors can deliver significant quantities of nutrients, i.e., phosphorus and nitrogen, from oceans to insular islands. On the other hand, they also transport a large number of toxic contaminants, typically heavy metals and persistent organic matters (POPs), via food chains and biomagnifications, bringing about potential ecological risks (Allaway and Ashford 1984; Wendy and Polis 1999; Blais et al. 2005). Seabird population changes, foraging behaviors and dietary compositions are essential for seabird ecology study. To date, the relationship between seabird ecology and climatic/environmental changes remains an open question and has not been well-studied. Seabird relics provide valuable information of the past, and thus enable a potential use to study changes in seabird ecology, paleoceanography and palaeoclimate. Elemental geochemical and isotopic analyses of the ornithogenic sediments from the Xisha Islands, are extensively used in the present study. From a technical perspective, this will aid in development in stable isotope ecology and island ecological geology. It will further helpful for examining biogeochemical cycling of some elements among geosphere, hydrosphere, biosphere and atmosphere. Under a background of global warming, our study constructs the basis of assessing the vulnerability of coral islands, as well as environmental management and resource exploitation.

According to the 4th synthesized report of IPCC in 2007, there is regional imbalance among countries. Most of reconstructed records are from developed countries, but data from developing countries are extremely deficient. The study of the Xisha Islands helps in understanding teleconnections among climate, oceans and creatures in low-latitude tropical areas, and can provide supplementary data from developing countries.

## 2.2 Research Objectives

Based on rationale of ecological geology, this study focuses on ecological responses of seabirds from a specific region. Study area of the present study is the Xisha archipelago located in the central South China Sea. We will attempt to investigate palaeoecology from ancient bird remains by elemental geochemical, stable isotopic and biochemical analyses. Seabird ecology in the past of this area is the core of the present study, and the principal objectives are specified as follows:

1. By high-resolution $^{210}Pb$-$^{137}Cs$ and radiocarbon ($^{14}C$) dating techniques, establish reliable chronologies for the sediment cores collected from five individual islands.
2. Via biomarker analysis (bio-elements and reflectance spectroscopy), reconstruct relative seabird population size over the past two millennia on the Xisha Islands, identify the overall trends in seabird population dynamics, and construct evolution model of coral island ecosystem.
3. From proxy-based analysis (including heavy metal mercury and black carbon, which reflects human metallurgy civilization and energy structure changes, respectively), seek for possible evidence of past anthropogenic activity.
4. Through stable isotopic analysis of ancient bird remains, reconstruct seabird palaeo-diet and examine interactions among seabird diet, foraging behavior and population changes.
5. From a historical perspective, demonstrate impacts of climate, ocean productivity and human activity on seabird population, investigate the reason for the

rapid decrease in seabird population in recent times, and assess ecological risks from human activities on the coral islands.

The ultimate objective of our study is providing useful data for future prediction of ecological response to climate change, on the basis of regional comparison.

## 2.3 Research Contents

1. Establishment of chronology for the collected sediment cores
   Accurate and reliable chronology is the basis of palaeoecology study. There are several kinds of dating techniques to choose from, depending on the type of material and its age. Because the Xisha Islands were formed during middle to late Holocene (~6 ka), and there are lots of carbon-bearing biological materials (bird/fish bone, eggshell etc.) in the sediments, we attempt to take radiocarbon analysis as the main dating technique (upper limit ~50 ka years). To examine possible influences of recent human activity on the island ecosystems, high-resolution chronology within the past ~150 years is indispensable. As radiocarbon ($^{14}C$) has a half-life approximately 5,730 years, it cannot well-resolved chronology of post-industrial sediments. For such deposits, $^{210}Pb$-$^{137}Cs$ dating is an ideal chronological analysis tool. Joint use of $^{14}C$ analysis and $^{210}Pb$-$^{137}Cs$ dating can yield more precise chronology. Detailed chronological analysis of the sediment cores will be given in following chapters.
2. Seabird population reconstruction
   Population is an important part of seabird ecology, and also a critical research topic in study of ecological responses to climate change and human activity. In an earlier study, record of seabird population since 1,300 years before present on Dongdao Island of the Xisha archipelago has been reconstructed. We have collected new sediment cores from the study area. On the basis of biomarker identification (avian bio-elements and characteristic spectrum), the present study attempt to reconstruct seabird population record over the past 2,000 years on the Xisha Islands. Based on bottom-up changes in source materials of the cores, we also infer the developmental model (i.e. gradual development) of such coral island. The influence of seabird activity on plant development and its feedback will also be discussed.
3. In-depth discussion of the causes for seabird population changes
   On the foundation of chronology analysis and seabird population reconstruction, our next step it to discuss the possible reasons for seabird population variability. Both natural and anthropogenic forcing can exert significant impacts on seabird abundance. Proxy-based climatic and environmental records have been reconstructed by scientists. These include historical SST, ENSO, monsoons and solar irradiation records, which may affect seabird directly or indirectly. We will compare our seabird population records with climatic dynamics and attempt to identify key factors affecting seabirds. It is expected to formulate

a theory explaining the mechanism of past seabird population changes in the Xisha Islands.

4. Reconstruction of seabird dietary compositions during the past 2,000 years
Generally, alterations in diets and foraging behavior of seabird lead to changes in stable isotope compositions. Thus, the palaeo-dietary record of seabirds can be obtained by stable isotope analysis of well-preserved bird remains, including organic matrix of eggshell, guano particle and bone collagen. As foraging strategy is closely related with population size, and seabird population changes may affect their dietary compositions. Potential linking among changes in climate, seabird foraging behavior and their population will be examined.
5. Records of past human activity from the ornithogenic sediments
Strength of human activity over the past 200 years has surpassed that of any other epoch. Heavy metal mercury and black carbon are closely related to human metallurgy civilization and energy structure. We will attempt to examine the levels of these proxies in ancient bird tissues, as well as bulk ornithogenic sediments. A potential correlation between anthropogenic activity and their levels will also be examined.
6. Preliminary study of ancient DNA extraction from subfossil guano particles
So far, little is known about the correlation between evolution and environmental changes on the Xisha Islands. Whether climate change impacts on evolution, or DNA structure, remains unclear. Bird droppings are remains of swallowed food. Such materials contain information about host organism, as well as prey. Ancient guano has a potential use to study temporal changes in DNA structure of both host and prey organisms. In the present, it will be attempted to perform DNA isolation from ancient guano particles.

## 2.4 Methodology

The overall idea of the present study is to clarify possible mechanism of seabird ecology changes in response to climate change and anthropogenic activity. We attempt to reconstruct records of seabird ecology in a tropical environment via biomarker identification. The basic principles of multi-disciplines, i.e. ecology, geology, oceanography, geochemistry etc., will be incorporated in the present study. Techniques that we employ include elemental geochemical, isotopic geochemical and biochemical analyses. "Macro" changes in seabird population/diet and "micro" alterations in trace elements and isotopic compositions will be evaluated by geochemical analyses. Proxy-based records and modern observations will also be compared to infer time-series evolution. Regional characteristics and differences of our records in tropical island ecosystems are also considered.

In short, our first step is to collect ornithogenic sediment cores from typical islands the Xisha archipelago. Then, accurate and reliable chronology of the sediment profiles will be established via radiocarbon and $^{210}Pb$ dating techniques.

After that, we will attempt to reconstruct developmental processes of the coral islands, seabird population and their dietary records over the past 2,000 years. Ultimately, it is expected to conceive a hypothesis explaining interactions among climate system, marine environments, seabirds and coral islands.

## References

Allaway WG, Ashford AE (1984) Nutrient input by seabirds to the forest on a coral island of the Great Barrier Reef. Mar Ecol Prog Ser 19:297–298

Blais JM, Kimpe LE, McMahon D, Keatley BE, Mallory ML, Douglas MSV (2005) Arctic seabirds transport marine-derived contaminants. Science 309:445

Guo ZT, Ruddiman WF, Hao QZ, Wu HB, Qiao YS, Zhu RX, Peng SZ, Wei JJ, Yuan BY, Liu TS (2002) Onset of Asian desertification by 22 Myr ago inferred from loess deposits in China. Nature 416:159–163

IPCC (2007) Climate Change 2007: Synthesis Report

Liu ZH, Herbert TD (2004) High-latitude influence on the eastern equatorial Pacific climate in the early Pleistocene epoch. Nature 427:720–723

Liu XD, Zhao SP, Sun LG, Luo HH, Yin XB, Xie ZQ, Wang YH, Liu KX, Hu XH, Ding XF, Fu DP (2006) Geochemical evidence for the variation of historical seabird population on Dongdao Island of the South China Sea. J Paleolimnol 36:259–279

National Research Council of the National Academies (2006) Surface temperature reconstructions for the last 2000 years. The National Academies Press, Washington, DC

Petit JR, Jouzel J, Raynaud D, Barkov NI, Barnola JM, Basile I, Bender M, Chappellaz J, Davis M, Delaygue G, Delmotte M, Kotlyakov VM, Legrand M, Lipenkov VY, Lorius C, PePin L, Ritz C, Saltzman E, Stievenard M (1999) Climate and atmospheric history of the past 420,000 years from the Vostok ice core, Antarctica. Nature 339:429–436

Shi YF (1997) Past 2000 years records and global change. Quat Sci 1:37–40

Sun LG, Xie ZQ, Zhao JL (2000) A 3000-year record of penguin population. Nature 407:808

Thompson LG, Yao T, Mosley-Thompson E, Davis ME, Henderson KA, Lin PN (2000) A high-resolution millennial record of the south Asian monsoon from Himalayan ice cores. Science 289:1916–1919

Wang PX (1999) Response of western pacific marginal seas to glacial cycls: paleoceanographic and sedimentological features. Mar Geol 156:5–39

Wang YJ, Cheng H, Edwards RL, Kong XG, Shao XH, Chen ST, Wu JY, Jiang XY, Wang XF, An ZS (2008) Millennial- and orbital-scale changes in the East Asian monsoon over the past 224,000 years. Nature 451:1090–1093

Wendy BA, Polis GA (1999) Nutrient fluxes from water to land: seabirds affect plant nutrient status on Gulf of California Islands. Oecologia 118:324–332

# Chapter 3
# Study Area and Sample Collection

## 3.1 Introduction to the South China Sea

### *3.1.1 Geological Settings and Natural Resources*

The South China Sea extends northwards from 0°N to approximately 20°N at the southern coast of China (Fig. 3.1). It is the largest marginal sea in the western Pacific and is sometimes referred as "Asian Mediterranean". It has an area of approximately 3.3 million km$^2$ (not including Thailand Gulf and Tonkin Gulf). The regions around it are densely populated (around 270 million), and major countries/regions include Malaysia, Indonesia, Philippines, Taiwan, Thailand and China. Significant rivers drain into it, e.g., Pearl River (in southern China), the Red River (in northern Vietnam), Mekong River (in southern Vietnam) etc. Important cities associated with this area include Guangzhou, Hong Kong, Hanoi and Saigon. The South China Sea is enriched in petroleum and natural gases. For example, potential reserve of China southern coast only is estimated to be about 1,500 million barrels. The reserve of gas and oil resources in South of Hainan Island is also above 200 million barrels. In addition, the South China Sea has also a high level of biodiversity. For example, 17 species of seagrass (globally 58 species), 3,365 species of fish, as well as a large number of mangroves, have been found in this area (Randall and Lim 2000). There are four species of horseshoe crabs worldwide, and three of them appear in Asia (the other one lives only on the east coast of the America). There are many dotted islands in the South China Sea, including Hainan Island, Taiwan Island, Luzon Island, Dongsha Islands, Zhongsha Islands, Xisha Islands, Nansha Islands etc. (Morton and Blackmore 2001). The depth of its central water is approximately 4,700 m, and the volume of the sea basin is around 4.24 million km$^3$. Its surface is characterized by warm water, low



L. Xu, *Impact of Climate Change and Human Activity on the Eco-environment*,
Springer Theses, DOI 10.1007/978-3-662-45003-1_3

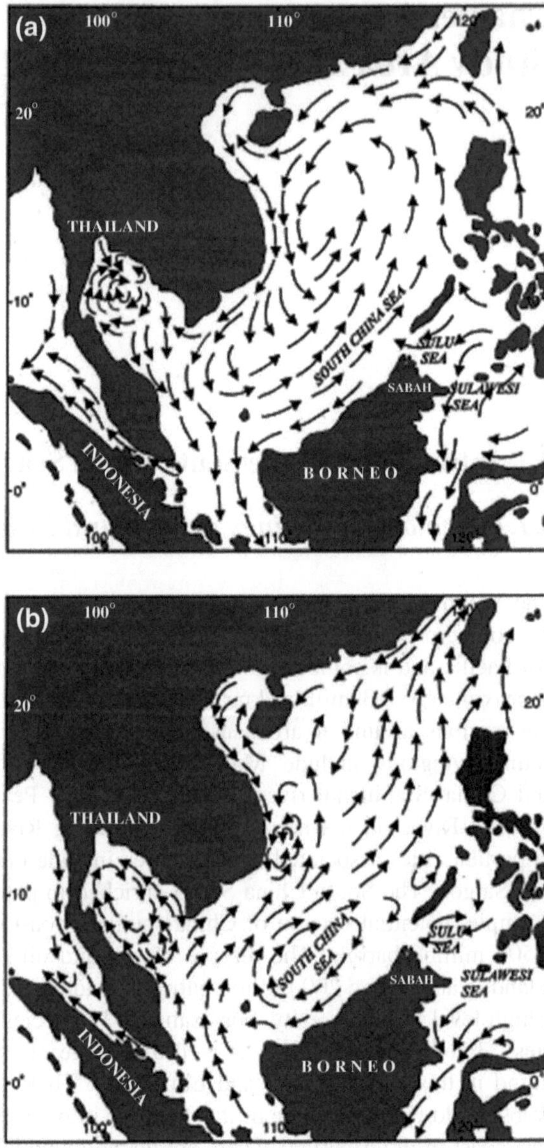


**Fig. 3.1** Seasonal surface circulation of the South China Sea: **a** winter and **b** summer (*Source* Morton and Blackmore 2001). [Reprinted from Morton and Blackmore (2001) Copyright (2001), with permission from Elsevier]

salinity, low nutrient availability and productivity. Coastal water of the South China Sea is quite shallow, normally less than 120 m. During the Last Glacial Maximum (~20 ka), sea level decreased at a magnitude of 100–120 m, making the surface circulation quite different from that of present. Drainage outlet at the southern part was totally closed then, and the Bashi Strait became its only connection to open sea (Wang et al. 1995).

### 3.1.2 Prevailing Monsoons and Ocean Circulation

The South China Sea has a tropical climate, and both its climate and ocean surface circulation are largely controlled by East Asian Monsoons. East Asian Monsoon is a branch of Asian Monsoon and driven by the thermal contrast between Asia and the tropical ocean. In boreal winter, southward East Asian Winter monsoon bring dry/cold air mass and move across mainland China. Northeast monsoons thus prevail over the South China Sea. In contrast, wet/warm and southwest summer monsoon moves from May to September, bringing heavy precipitation to Asia. On large time scales, it is generally accepted that strength of East Asian Winter Monsoon is negatively correlated with East Asian Summer Monson. Recently, Wang et al. (2011) identified that the correlation between East Asian Winter Monsoon and Summer Monsoon is variable, with dependence on the strength of East Asian Winter Monsoon. Under the influence of seasonal monsoons, an anti-clockwise pattern of surface circulation is created in boreal winter, and the surface current flow in the South China Sea is reversed in summer, with a few eddies (as shown in Fig. 3.1; Morton and Blackmore 2001). Major ocean currents in the South China Sea include Taiwan Current and Kuroshio Current. The cold Taiwan Current enters the South China Sea via Taiwan Strait. The Kuroshio Current is warm and runs into the South China Sea through Bashi Strait, reducing the cooling effect of Taiwan Current (Hu et al. 2000). Moreover, summer current from the Indian Ocean can significantly warm the South China Sea (Wang et al. 1995).

In addition to seasonal wind-driven surface circulation, coastal upwelling is also an important part of ocean circulation in the South China Sea. Seasonal monsoons induce seaward Ekman transport and further coastal upwelling (Ning et al. 2004). Upwelling is normally cold and has a relatively high salinity and high potential density. Nutrient-rich upwelling significantly impacts on primary production of the South China Sea. Plankton bloom changes the color of the ocean surface, and satellite-based ocean color observation provides evidence for the changes in ocean productivity (Kuo et al. 2000). There are typical upwelling regions off the South Vietnam, Luzon Island and northern China. South Vietnam upwelling in western South China Sea is controlled by summer monsoon, while the upwelling off the Luzon is dominated by winter monsoon (Jing et al. 2007; Xie et al. 2003; Shaw et al. 1996; Tang et al. 2011). Upwelling system in the northern continental shelf of the South China Sea refers to east coast of Hainan Island and Leizhou peninsular, and southeast of the Zhanjiang Bay (Jing et al. 2007). Vietnam coastal upwelling in summer and Luzon Island coastal upwelling in winter are two critical components of upwelling systems in the South China Sea. Their locations are shown as shaded areas in Fig. 3.2 (Jian et al. 2001). As those two regions are quite sensitive to East Asian Summer Monsoon and East Asian Winter Monsoon, marine sediments from these areas have been used to study the evolution of East Asian monsoon. For instance, Huang et al. (2002) analyzed stable isotope compositions in foraminifer samples from cores 17,928 and 17,954. They reconstructed dynamics of upwelling system over the past 220 ka years and found that the intensity of upwelling is closely related to monsoon

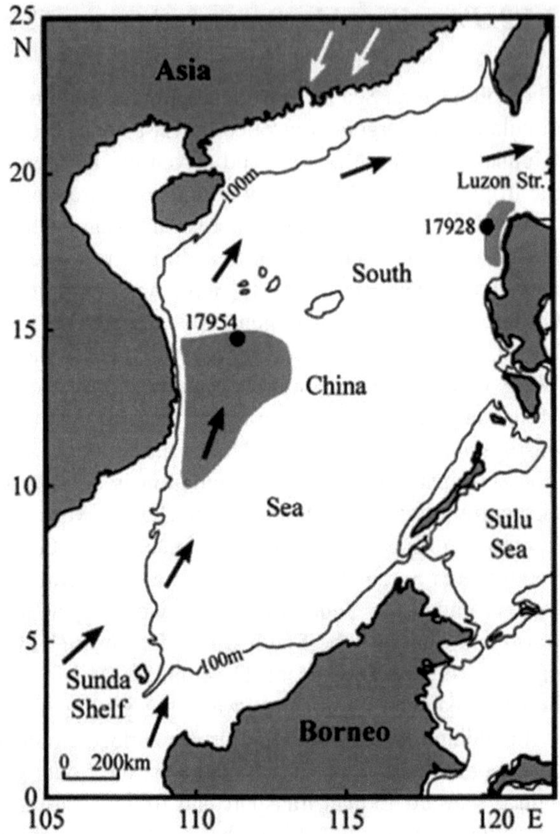


**Fig. 3.2** Modern upwelling areas in western and eastern South China Sea (*shaded areas*); *black* and *white arrows* show summer and winter monsoon, respectively (*Source* Jian et al. 2001). [Reprinted from Jian et al. (2001). Copyright (2001), with permission from Elsevier.]

strength during glacial-interglacial cycles. In general, both observations and simulation data show that productivity in winter is higher than that in summer, and the productivity across the whole basin is driven by winter monsoon (Liu et al. 2002). In the South China Sea, the overall upwelling regions account for a small part of the whole basin. Xisha archipelago is located neither the summer upwelling region, nor winter upwelling area. Which part of the East Asian monsoons, winter or summer, plays a dominant role in productivity of the Xisha Islands remains an open question.

## 3.2 Introduction to the Xisha Islands

### *3.2.1 Geological Settings of the Xisha Islands*

Ths Xisha Islands (also known as Xisha archipelago, Paracel Islands) is located in the central South China Sea and approximately 182 miles away from Sanya City, Hainan province (15°47′–17°08′N, 110°10′–112°55′E, Fig. 3.3a). It consists of

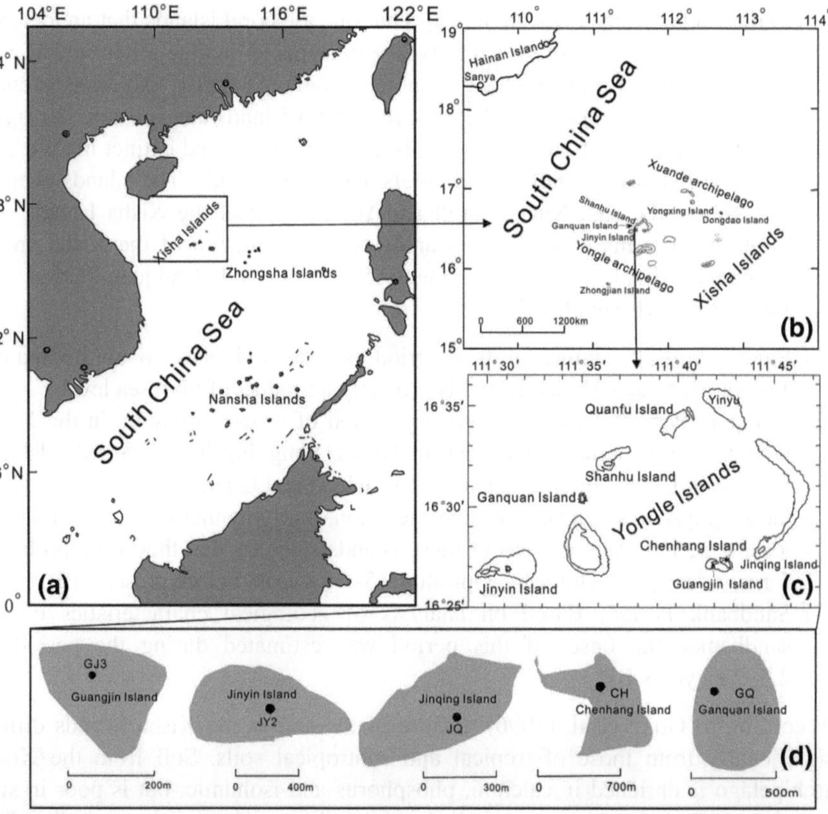


**Fig. 3.3** Study area and sampling sites **a** South China Sea; **b** Xisha Islands; **c** Yongle Islands; **d** Guangjin Island, Jinyin Island, Jinqing Island, Chenhang Island and Ganquan Island showing sampling sites

more than 40 islets, sandbanks and reefs. The archipelagic water and land areas cover ~500,000 and 8 km$^2$ (Fig. 3.3b), respectively. Along the northeast-southwest direction, The Xisha Islands can be divided into two groups: the eastern and the southwestern islands (Fig. 3.3c). The Xuande archipelago in the east consists of Zhaoshu Island, North Island, Middle Island, South Island, Yongxing Island, Shidao Island, Dongdao Island and many sand cays. The southwestern group, the Yongle archipelago, is composed of Shanhu Island, Ganquan Island, Jinyin Island, Guangjin Island, Jinqing Island, Chenhang Island and Zhongjian Island, as well as sandbanks and reefs (Hainan Ocean Administration 1999). According to observations from Yongxing meteorological station, annual mean air temperature and annual rainfall of the Xisha Islands are 26–27 °C and 1,500 mm, respectively. From June to November, the Xisha Islands are subject to the effects of the southwest monsoon, frequent tropical cyclones arising from intense convergent convection, and heavy precipitation accounting for more than 80 % of the annual total (Lin et al. 1999; Hainan Ocean Administration 1999).

All the islets of the Xisha archipelago but one are coral islands that are derived from coral sand debris. Corals can only live in a special marine environment, and are quite sensitive to small changes in temperature, salinity and transparence. According to Charles Darwin's theory, ocean coral islands are volcanic in origin. At the initial phase, reef building corals began to grow around extinct marine volcanos and ultimately "manufactured" reefs and coral islands. The islands eventually become atolls, e.g., Xuande atoll and Yongle atoll of the Xisha Islands (Lu et al. 1979). In terms of time-series analysis, the formation of the Xisha archipelago can be divided into three stages: Shidao Period, Ganquan Period and Sandbank Period (Lu et al. 1979).

(1) Shidao Period: Timing of this period is extended back to approximately 120 ka years, and Shidao Island is probably a product of high sea level.
(2) Ganquan Period: This is the primary period of island formation in the Xisha archipelago. Ganquan Island, North Island, Yongxing Island, Shanhu Island, Chenhang Island, Jinqing Island, Jinyin Island and Guangjin Island have similar geophysical settings, as well as elevation, implying that they formed during a same period. Age analysis of these islands suggests that they were probably formed during the Holocene Optimum (5–6 ka years before present, BP).
(3) Sandbank Period: Based on analysis of geological characteristics of the sandbanks, the onset of this period was estimated during the period of 2.5–2 ka years BP.

According to Gong et al. (1996), texture of the soil in the Xisha Islands differs significantly from those of tropical and subtropical soils. Soil from the Xisha archipelago is enriched in calcium, phosphorus and isohumic, but is poor in silicon, iron and clay, and has a relatively high salinity. The soil is normally alkaline and has a pH around 8–9. Under the influence of pedogenesis materials (coral sand), the impacts of soil structure on vegetation development have surpassed that of climate. Tropical rainforest is thus unable to develop on the Xisha Islands (Exploration Group of Xisha Islands of Plant Institute of Guangdong Province 1977).

### *3.2.2 Flora and Fauna on the Xisha Islands*

Due to the long distance from mainland China and its special soil structure, plant taxa on the Xisha archipelago are relatively simple. According to the investigations into vegetation on the Xisha Islands, a total of 213 plant species (57 families and 154 genera, including cultivated plants) were found (Exploration Group of Xisha Islands of Plant Institute of Guangdong Province 1977). 166 of the 213 plant species are wild and most of them are typical tropical/subtropical plants, e.g., *Mimosaceaae* and *Caesalpiniaceae*. Due to a low elevation and small areas of the islands, coastal and insular plants were also observed on the Xisha archipelago, including *Cassythafiliformis*, *Sesuvium portulacastrum* etc. *Angiopteris fokiensis* and *phymatodes scolopendria* are the only two pristine pteridophytes in the Xisha Islands. In terms of vegetation type, the plants on the Xisha Islands can be divided into five groups. (1) Indeciduous arbors, e.g., *Pisonia grandis, Cuettarda speciosa*; (2) Indeciduous shrubs, e.g., *Scaevola scricea, Messerschmidia argentea, Pemps acidula*; (3) Tropical herbaceous plants, e.g., *Thuarea involuta, Ipomoea pes-caprae, Eragrostis ciliate*; (4) Limnophyte communities, including *Sesuvium portulacastrum, Ruppia rostellata* etc.; (5) Cultivated plants, most of whom are edible, e.g., vegetables and fruits.

Most of the Xisha islands are covered with thriving vegetation, and typical plant communities display circular-zonary growth patterns around the islets. The interiors of most of islets are covered by trees such as *Pisonia grandis* and *Guettarda speciosa*, and bordered by shrubs such as *Scaevola sericea, Messerschmidia argentea* and *Morinda citrifolia*. The abundant vegetation provides a good habitat for seabirds. There were once more than 60 species of birds on these islands, resulting in the accumulation of high levels of guano in the soil (Exploration Group of Xisha Islands of Institute of Soil Science of Chinese Academy of Sciences 1977).

Although the number of species on the Xisha Islands is smaller than that on continents and continental islands, there were a large number of seabirds inhabiting these insular islets, including boobies, terns, frigatebird etc. Among those birds, Red-footed Booby (*Sula sula*) and Brown Booby (*Sula leucogaster*) are the dominant species. There are also mammals, i.e. mice and introduced species cattle, on Dongdao and Yongxing islands. Due to its long distance from mainland China and restrictions imposed by the Chinese military on travel to this area, the Xisha archipelago remains in a relatively pristine condition. Natural environmental archives in the Xisha archipelago have been well preserved and thus provide an ideal window for observing and predicting global environmental changes (Zhao 1996).

## 3.3 Sample Collection

The pristin environment of the Xisha archipelago has been well-preserved and it is an ideal place to study developmental history of tropical coral island ecosystem in the ocean, but our knowledge of past eco-environmental changes on these insular islands is limited. So far, the long-term interaction between plant development and seabird activity on these islands of the South China Sea has not been well documented.

To obtain a better understanding of the ecology and bird activities in the Xisha archipelago, we performed systematic investigations into the ornithogenic sediments on the Xisha atoll, in 2003 and 2008. Environmental medium samples (including plant, coral sand, fresh guano) and several sediment cores were collected during field studies. Seabird ecology on five islands of the Xisha archipelago will be extensively discussed in the present study (Fig. 3.3c). Ganquan

Island (GQ), Guangjin Island (GJ3), Jinqing Island (JQ), Chenhang Island (CH) and Jinyin Island (JY2) are all coral islets developed from the Yongle atoll. They were formed during the middle to late Holocene, with an area of 0.3, 0.06, 0.21, 0.36 and 0.33 $km^2$ (Fig. 3.3d), respectively. As stated above, the characteristic landscape of these islands is circular-zonary growth of plants around the islets (Exploration Group of Xisha Islands of Institute of Soil Science of Chinese Academy of Sciences 1977; Zhang 1974). All the sediment cores used in this study, except for core JY2 (collected during an earlier expedition in 2003), were collected during an expedition in 2008. Sediment cores GQ (16°30′15.1″N, 11°35′6.2"E), GJ3 (16°27′07″N, 111°42′5″E), JQ (JQ, 16°27′50″ N, 111°44′27″ E), JY2 (16°26′57″N, 111°30′24″E) and CH (16°27′10.1″N, 111°42′34.0″E) were collected under woodland and shrubs on Ganquan, Guangjin, Jinqing, Jinyin and Chenhang Islands, respectively (Fig. 3.3d). When sampling, we inserted 11 cm diameter PVC plastic gravity pipes of lengths 107 cm (GQ), 55 cm (JY2), 95 cm (GJ3), 55 cm (JQ) and 72 cm (CH) into the soft substrate and then excavated sediments around the pipes to retrieve those pipes (Figs. 3.4, 3.5). An overview of these five islands and sediment cores is given in Table 3.1. Duplicate 1 m × 1 m pits were also dug, and color, grain size and visible components of the sediments were described in detail. Meanwhile, the coarse fraction of each duplicate sediment sample was separated using a 10-mesh stainless steel sieve to obtain sufficient guano particles, and avian bones and other biological remains in situ (Fig. 3.5). The loose sediments were a mixture of humus, guano and coral sand, with a few well-preserved remains of bird and fish bones, and fish scales. Those cores had a similar lithology. Large quantities of black humus were observed in the upper layers, likely indicating a high level of organic matter (OM). The color changed gradually from black to light yellow down the cores. Large numbers of guano pellets were observed overall, but the bottom layers consisted primarily of medium-large size coral sands with few yellow guano particles, reflecting a remarkable decrease in OM.

**Fig. 3.4** In situ sampling sites during our second expedition into the Xisha Islands in 2008, sediment cores JQ, GJ3, GQ, CH and JY2 (not shown) were collected from Jinqing Island, Guangjin Island, Ganquan Island, Chenhang Island and Jinyin Island, respectively

**Fig. 3.5** Photos showing sediment core (*left panel*) and bird remain collection (*right panel*) from duplicate profiles

**Table 3.1** Parameters of the studied islands and sediment profiles

| Island name | Sampling site | Elevation (m) | Length × Width (m × m) | Area ($km^2$) | Core code | Core length (cm) |
|---|---|---|---|---|---|---|
| Jinyin Island | 16°26′57″N, 111°30′24″E | 0.2 | 1020 × 350 | 0.36 | JY2 | 55 |
| Guangjin Island | 16°27′07″N, 111°42′5″E | 6–8 | 320 × 200 | 0.06 | GJ3 | 95 |
| Jinqing Island | 16°27′50″ N, 111°44′27″ E | 6.0 | 880 × 230 | 0.20 | JQ | 55 |
| Ganquan Island | 16°30′15.1″N, 11°35′6.2″E | 0.2 | 1020 × 350 | 0.3 | GQ | 107 |
| Chenhang Island | 16°27′10.1″N, 111°42′34.0″E | 5 | 315 × 1,050 | 0.33 | CH | 72 |

*Source* Haninan Ocean Administration (1999)

## 3.4 Preliminary Treatment of Samples

Prior to chemical analysis, the collected samples were pretreated as follows. These cores were opened, photographed, described and then sectioned at 1 cm intervals in the laboratory. An aliquot of each bulk sediments (approximately 10 g) was dried, and then homogenized with a pestle and mortar after being dried to a constant weight at a temperature of 105 °C, and were then passed through a 200 mesh sieve. Biological remains collected in the field were further separated with toothpick and stainless steel tweezer in the laboratory. Figure 3.6 shows some of the bird remains, including guano particles, bones and eggshells. As the Xisha archipelago is perennially dry due to a relatively high temperature and evaporation rate, the guano is thus dehydrated rapidly and can well-preserved for quite a long time.

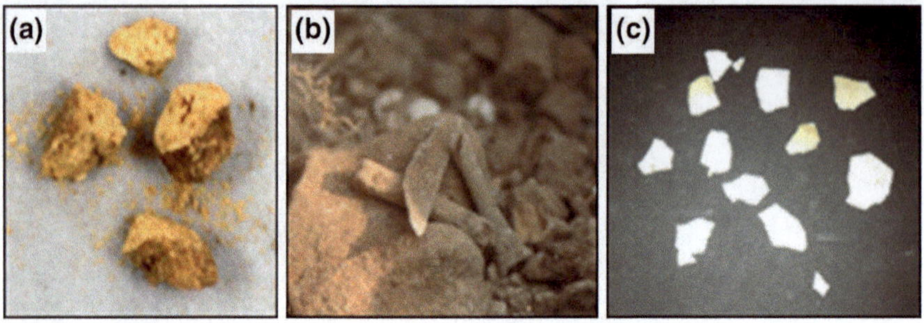


**Fig. 3.6** Biological remains collected from the ornithogenic sediments in the laboratory, **a** guano particle; **b** bird bone; **c** eggshells

## 3.5 Analytical Methods

We analyzed major and trace elements in the bulk sediments, as well as guano particles. For elemental analysis, approximately 0.25 g of each powdered sample was taken, precisely weighted and then digested by multi-acids ($HClO_4$–$HNO_3$–HCl–HF) in an electrically heated crucible. Following the detailed analysis methods reported in our previous studies (Liu et al. 2006, 2008), the concentrations of P, Cu, Cd, Zn, Ba, Fe, Al, K, Ti, Na, Mg, Sr, Ti, Ni and Mn were determined by inductively coupled plasma-optical emission spectrometry (ICP-OES, Perkin Elmer 2100DV). Arsenic and selenium levels were measured by atomic fluorescent spectrometry (AFS-930, Titan Instruments Co., Ltd.). Reagent blanks and certified reference materials (GBW07108/GBW07303) were used as quality assurance (QA) quality control (QC) in the analysis. The analytical precisions (RSD) for major and trace elements were better than ±0.5 and ±5 %, respectively. Total nitrogen (TN), total carbon (TC) and total hydrogen (TH) were determined by an elemental analyzer (Vario EL III) with a precision (relative standard deviation, RSD) better than 1 %. The chemical volumetric method was used to measure total organic carbon (TOC) within a RSD of 0.5 %, and the $CaCO_3$ level was determined by subtracting the TOC from the total carbon (Chang et al. 2005; Bernárdez et al. 2008). The analytical methods of elements are summarized in Table 3.2.

Except for above-mentioned elemental analysis, the present study refers to multi-proxy analyses, including test of some radionuclides (i.e. $^{210}Pb$, $^{137}Cs$),

**Table 3.2** Analytical methods, instrumentations in the present study

| Elements | Equipments/methods | Reference standards |
|---|---|---|
| Cu Zn Cd As Se P Ba Pb $Fe_2O_3$ Al $K_2O$ $Na_2O$ MgO Sr Mn Ti Ni | ICP-AES | GBW07108 |
| As, Se | AFS-930 | GBW07108/GBW07303 |
| TOC | Volumetric method | GBW07108 |
| TC, TN, TH | Elemental analyzer | Sulfadiazine |

Accelerated Mass Spectrometer analysis of ancient bones, black carbon and mercury analysis etc. For convenience of discussion, the analytical methods of such proxies will be given in individual chapters.

## References


Bernárdez P, González-Álvarez R, Francés G, Prego R, Bárcena MA, Romero OE (2008) Late Holocene history of the rainfall in the NW Iberian peninsula—Evidence from a marine record. J Mar Syst 72:366–382

Chang YP, Wu SM, Wei KY, Murayama M, Kawahata H, Chen MT (2005) Foraminiferal oxygen isotope stratigraphy and high-resolution organic carbon, carbonate records from the okinawa trough (IMAGES MD012404 and ODP Site 1202). Terr Atmos Oceanic Sci 16(1):57–73

Exploration Group of Xisha Islands of Institute of Soil Science of Chinese Academy of Sciences (1977) Soil and guano phosphorus mine in Xi-Sha Islands. Science Press, Beijing (in Chinese with English abstract)

Exploration Group of Xisha Islands of Plant Institute of Guangdong province (1977) Plants and vegetation on Xisha Islands of China. Science Press, Beijing (in Chinese with English abstract)

Gong ZT, Liu LW, Zhou RR (1996) Formation of the soils in the South China Sea islands and their ages. Quat Sci 1:88–95 (in Chinese with English abstract)

Hainan Ocean Administration (1999) The professional proceedings of the integrated investigation research on sea islands resource of Hainan province. China Ocean Press, Beijing (in Chinese with English abstract)

Hu JY, Kawamura H, Hong HS, Qi YQ (2000) A review on the currents in the South China Sea: seasonal circulation, South China Sea warm current and Kuroshio intrusion. J Oceanogr 57:607–624

Huang BQ, Jian ZM, Cheng XR, Wang PX (2002) Foraminiferal responses to upwelling variations in the South China Sea over the last 220,000 years. Mar Micropaleontol 47:1–15

Jian ZM, Huang BQ, Kuhnt W, Lin HL (2001) Late quaternary upwelling intensity and East Asian monsoon forcing in the South China Sea. Quatern Res 55:363–370

Jing ZY, Hua ZL, Qi YQ, Zhang H (2007) Summer upwelling in the northern continental shelf of the South China Sea. 16th Australasian fluid mechanics conference. Crown Plaza, Gold Coast. 2–7 Dec 2007

Kuo NJ, Zheng QN, Ho CR (2000) Satellite observation of upwelling along the western coast of the South China Sea. Remote Sens Environ 74:463–470

Lin X, Chen XL, Wang F (1999) The report on investigation of sea island of Hainan province. China Ocean Press, Beijing (in Chinese with English abstract)

Liu KK, Chao SY, Shaw PT, Gong GC, Chen CC, Tang TY (2002) Monsoon-forced chlorophyll distribution and primary production in the South China Sea: observations and a numerical study. Deep Sea Res I 49:1387–1412

Liu XD, Zhao SP, Sun LG, Luo HH, Yin XB, Xie ZQ, Wang YH, Liu KX, Hu XH, Ding XF, Fu DP (2006) Geochemical evidence for the variation of historical seabird population on Dongdao Island of the South China Sea. J Paleolimnol 36:259–279

Liu XD, Zhao SP, Sun LG, Yin XB, Xie ZQ, Wang YH (2008) Paleoenvironmental implications of the Guano phosphatic cementation of coral calcarenite on the Dongdao Island in the South China Sea. Mar Geol 247:1–16

Lu YC, Yang XC, Jia RF (1979) Quarternary biological sediments in the Xisha Archipelago, China and a discussion on the age of island-formation. Geochimica 2:93–102 (in Chinese with English abstract)

Morton B, Blackmore G (2001) South China Sea. Mar Pollut Bull 42:1236–1263

Ning X, Chai F, Xue H, Cai Y, Liu C, Shi J (2004) Physical-biological oceanographic coupling influencing phytoplankton and primary production in the South China Sea. J Geophys Res 109:C10005. doi:10.1029/2004JC002365
Randall JE, Lim KKP (2000) A checklist of the fishes of the South China Sea. Raffles Bull Zool Suppl 8:569–667
Shaw PT, Chao SY, Liu KK, Pai SC, Liu CT (1996) Winter upwelling off Luzon in the northeastern South China Sea. J Geophys Res 101(C7):16435–16448
Tang SL, Dong Q, Liu FF (2011) Climate-driven chlorophyll-a concentration interannual variability in the South China Sea. Theor Appl Climatol 103:229–137
Wang PX, Wang LJ, Bian YH, Jian ZM (1995) Late Quaternary paleoceanography of the South China Sea: surface circulation and carbonate cycle. Mar Geol 127:145–165
Wang L, Li JJ, Lu HY, Gu ZY, Rioual P, Hao QZ, Mackay AW, Jiang WY, Cai BG, Xu B, Han JT, Chu GQ (2011) The East Asian winter monsoon over the last 15,000 years: its links to high-latitudes and tropical climate systems and complex correlation to the summer monsoon. Quatern Sci Rev 32:131–142
Xie SP, Xie Q, Wang DX, Liu WT (2003) Summer upwelling in the South China Sea and its role in regional climate variations. J Geophys Res 108(C8):326. doi:10.1029/2003JC001867
Zhang HD (1974) The vegetation of the His-Sha Islands. Acta Botanica Sinica 16(3):183–190
Zhao HT (1996) History of expeditions to Xisha Islands. Geogr Res 15(4):55–65 (in Chinese with English abstract)

# Chapter 4
# Chronology

Chronology, or dating, is a prerequisite in palaeoclimate and palaeoecology studies. There are several typical techniques to dating natural archives (e.g. lacustrine/marine sediments, tree rings, stalagmite, ice cores, loess etc.). These methods include chemical analysis (e.g. amino acid dating), palaeomagnetic dating, dendrochronology and radiometric dating. Of these approaches, radiometric dating is one of the mostly used dating techniques. The radiometric dating methods used in the present study include $^{210}Pb$ and radiocarbon ($^{14}C$) chronology analyses. Radiocarbon, i.e. $^{14}C$, is an ideal radionuclide for studying the ages of samples deposited within the Holocene (~10 ka). The widely known and accepted half-life of $^{14}C$ is 5,730 years, thus it can well determine ages of carbon-bearing ancient materials from hundreds to ~50 ka years old. Chronology within the past 200 years had not bee well documented until successful application of $^{210}Pb$ in 1963 to date recent sediments (Goldberg 1963).

## 4.1 Age Determination of Coral Sand Ornithogenic Sediments by High-Resolution $^{210}Pb$-$^{137}Cs$ Dating

### *4.1.1 About $^{210}Pb$ Dating*

In the $^{238}U$ decay series, radiometric $^{210}Pb$ is one of the natural daughter isotopes of its long-lived precursor $^{226}Ra$ (Fig. 4.1). Of all the $^{226}Ra$ daughter isotopes, $^{210}Pb$ is the only long-lived daughter radionuclide (Half-life $T_{1/2} = 22.3$ years). Disequilibrium between $^{210}Pb$ and its parent isotope radium, i.e. $^{226}Ra$, in the series arises through diffusion of the intermediate gaseous isotope radon ($^{222}Rn$). Part of the $^{222}Rn$ atoms resulted from $^{226}Ra$ in soils or sediments escapes into the atmosphere where they decay through a series of short-lived radionuclides to $^{210}Pb$, which is so called excess $^{210}Pb$ ($^{210}Pb_{ex}$). The excess $^{210}Pb$ is brought back



L. Xu, *Impact of Climate Change and Human Activity on the Eco-environment*, Springer Theses, DOI 10.1007/978-3-662-45003-1_4

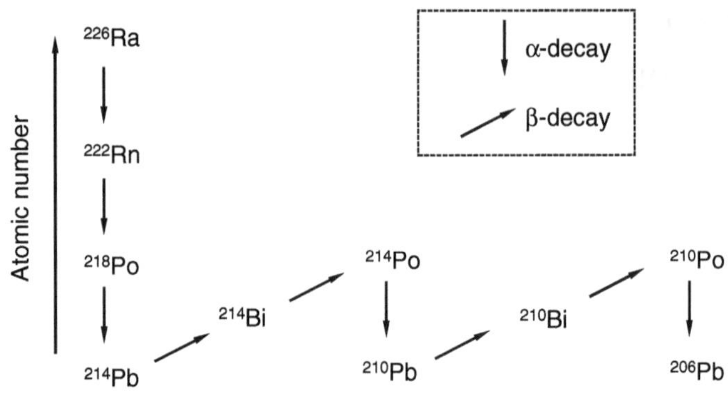


**Fig. 4.1** Decay chain products of long-lived $^{226}$Ra

to the earth surface by dry or wet precipitation (Appleby 2001). The decay process of excess $^{210}$Pb in the sediments then conforms strictly to the radioactive decay law. Excess $^{210}$Pb radioactivities are determined by subtracting the supported activity (supported $^{210}$Pb), which is in equilibrium with the in situ $^{226}$Ra in the sediments, from the total activities (Wan 1997). Half-life of radionuclide $^{210}$Pb is 22.3 years and, theoretically, can identify a range of ~200 years, linking recent sediments and older deposits that can be dated by $^{14}$C. At the early years of $^{210}$Pb dating, scientists questioned its reliability, but the introduction of reference time-marker $^{137}$Cs validated its high confidence. Independent dating technique is based on accumulation peaks of anthropogenic radiometric isotopes, typically $^{137}$Cs (half-life ~30 years) and $^{241}$Am, in sediments. $^{137}$Cs is an artificial radionuclide and it was accumulated to detectable level in 1954. The level of $^{137}$Cs reached its highest level in the year 1963 due to large scale human nuclear weapon test. A $^{137}$Cs peak in 1986, caused by the Chernobyl nuclear accident, has also been widely accepted. The joint use of $^{210}$Pb and $^{137}$Cs thus promoted development of young geological chronology.

Milestone events of $^{210}$Pb-$^{137}$Cs dating are listed as follows. $^{210}$Pb was successfully used to date ice core in 1963 (Goldberg 1963); it was then applied to lacustrine sediments in 1971 (Krishnaswamy et al. 1971); and Appleby and Oldfield (1978) further improved the dating models. The premise of $^{210}$Pb dating includes: (1) Sediments need to be within a geochemical closed system and sedimentation rate remains relatively stable; (2) $^{210}$Pb has a short residence time in water and thus can be transferred into sediments rapidly; (3) There is no significant migration of $^{210}$Pb after its deposition; (4) sediment sequence remains undisturbed.

Over the past several decades, numerous studies have shown that $^{210}$Pb dating could be successfully used for lake and marine sediments, as well as peat bog samples (Appleby 2008). However, it remains unclear whether this technique is applicable for dating ornithogenic sediments that are predominantly composed of coral sands and guano particles. Such sediments are widely distributed on the small islands of the Xisha archipelago, but in terms of structure, texture and geochemical compositions, they differ significantly from lake and marine sediments.

In the present study, the radioactivities of several radionuclides, including $^{226}$Ra, $^{210}$Pb and $^{137}$Cs, in five ornithogenic coral sand sediment cores (Refer to Chap. 3 for details) collected from different islands of the Xisha archipelago were determined. We aim to investigate the temporal and spatial distribution of radionuclides, and the potential driving factors, the influence of seabird activity and guano input on radionuclide distribution, and examine the applicability of the $^{210}$Pb dating technique for such ornithogenic coral sediments.

### 4.1.2 Analytical Methods

The radioactivity of $^{210}$Pb, $^{226}$Ra, $^{137}$Cs in our subsamples was measured by direct gamma spectrometry using Ortec HPGe GWL series, low background, coaxial, well-type, and intrinsic germanium detector. The detector, manufactured by AMETEK Co. Ltd, was placed inside a shield with 2 mm copper and 10 cm lead wall to effectively reduce the background radiation. The precision for $^{60}$Co at 1,332 keV is ±2.10 keV. The standard radionuclides i.e. $^{210}$Pb, $^{226}$Ra, $^{137}$Cs and $^{241}$Am, were provided and calibrated by China Institute of Atomic Energy. Radioactive isotopes $^{137}$Cs, $^{241}$Am, $^{155}$Eu, $^{57}$Co, $^{54}$Mn, $^{65}$Zn and $^{60}$Co were used in equipment calibration.

Prior to analysis, samples were processed as follows: they were first dried to constant weight at a temperature of 105 °C, then homogenized using a pestle and mortar, and further passed through a 10 mesh sieve. Samples were packed into standard counting geometries for gamma analyses. An 5–10 g aliquto of the ornithogenic sample was sealed and stored for about three weeks to allow radioactive equilibration between $^{214}$Pb and its parent isotope $^{226}$Ra. Energy spectra were obtained at a count time of approximately 24 h to obtain enough counts. The resulting spectrum files showed evident $^{210}$Pb peak at 46.5 keV. The $^{226}$Ra activity was measured by gamma-ray emissions of its daughter isotope $^{214}$Pb at 295 keV in the spectra, and $^{137}$Cs level was determined by its emission at 662 keV.

### 4.1.3 Distribution and Accumulation of Radionuclides in the Xisha Islands

There are no large animals living on these islands in the present day. Field observation showed no disturbance evidence for the sampling sites by direct human activities. At present, the Xisha region is still at a relatively pristine condition, and the collected sediments are thus well-preserved. To further examine whether these sediments constitute genuine environmental archives, we also determined the Accelerator Mass Spectrometry (AMS) $^{14}$C ages of several ancient bone samples in these sediments and we did not observe any age inversion.

The levels of dry bulk density and the activity concentrations of $^{210}$Pb, $^{226}$Ra, $^{137}$Cs, as well as the concentrations of Total Organic Carbon (TOC) and phosphorus in the subsamples of cores GQ, GJ3, JQ1 (duplicate profile of JQ) and JY2 are given in Table 4.1. The equilibrium between $^{210}$Pb and $^{226}$Ra in the bulk sediments is reached at depths of 7, 9, 9, 22.5 and 11 cm in GQ, GJ3, JQ1, CH and JY2, respectively (Fig. 4.2). The excess $^{210}$Pb activity concentration of GQ versus shows a simple exponential pattern, and the chronology of this core could be calculated using the Constant Initial Concentration (CIC) model (Appleby 2001). The excess $^{210}$Pb activities of JY2, GJ3, JQ1, and CH, however, do not decrease with depth in an exponential style. As discussed above, these sediments of each core are from a genuine natural sequence, thus the non-exponential pattern of excess $^{210}$Pb is possibly caused by changes in sedimentation rate. In general, the non-exponential pattern of excess $^{210}$Pb is acceptable for the Constant Rate of Supply (CRS) model (Appleby 2001). To examine whether $^{210}$Pb could be used for dating sediments from these four cores, we attempted to use CRS model for age calculation.

Chronologies of these five profiles and vertical distributions of $^{137}$Cs activities were provided in Fig. 4.2. In comparison with the reported $^{137}$Cs activity concentrations in lacustrine sediments from mainland China (Zhang 2005), the $^{137}$Cs activity concentration in our five cores is much lower. Nontheless, it still shows evident activity peaks. For example, both GQ and GJ3 profiles show well-defined single $^{137}$Cs peaks, corresponding to the 1963 fallout maximum. In contrast, the $^{137}$Cs records of JQ1 and JY2 have two distinct peaks. The older peaks were identified as a record of the 1963 fallout maximum, and the younger peaks might be a record of the Chernobyl nuclear accident in 1986. It seems that CH profile only recorded the 1986 peak. The absence of the younger peaks in GQ and GJ3 may be attributed to the different geochemical characteristics; $^{137}$Cs is more active than $^{210}$Pb (Xiang 2000). Except for JY2, the age of 1963 as derived from the $^{137}$Cs distribution pattern in the profiles of GQ, GJ3, JQ1 and CH is consistent with that from $^{210}$Pb dating. The ages for the sediments at the depths 2–3 cm (GQ), 5–6 cm (GJ3), 4–5 cm (JQ1) and 8–11 cm (CH) are 34–51 years, 37–51 years, 21–46 years and 20–22 years old, respectively. The $^{137}$Cs peaks of the year 1963 and 1986 are in the corresponding sediments of these three profiles (Fig. 4.2). The discrepancy between the ages from $^{210}$Pb and $^{137}$Cs for JY2 is probably resulted from downward diffusion of $^{137}$Cs. In addition, the sedimentation rates estimated from radiocarbon ages of the bone samples in the deeper layers of these sediments are parallel with the that from $^{210}$Pb ages. Such evidence suggests the reliability of the $^{210}$Pb dating method for the coral sand sediments enriched with guano in the Xisha islands. Figure 4.2 also shows that the sediments from surface layers of these profiles generally have a time span of approximately 150 years, rending those sediments ideal archives for studying sedimentation and ecological processes after the Industrial Revolution.

The inventories ($I_s$) of $^{210}$Pb measure its cumulative fallout and are frequently used to estimate the average $^{210}$Pb supply rate (or flux). $^{210}$Pb inventories were calculated based on dry bulk densities and radioactivities shown in Table 4.1. $I_s$ values for GQ, JQ1, GJ3 and JY2 are 3,628, 4,479, 3,468 and 4,583 Bq m$^{-2}$, respectively. As dry bulk density of CH sediments was not determined, the $I_s$ of $^{210}$Pb in this core is not given here. The equation $F = I_s\ \lambda$ is used to describe the simple relationship between excess $^{210}$Pb inventory and $^{210}$Pb supply rate

**Table 4.1** Activities of $^{210}Pb$, $^{226}Ra$, $^{137}Cs$, P and TOC in the ornithogenic coral sand sediments of Ganquan, Guangjin, Jinqing and Jinyin Islands

| Sample No | Depth (cm) | Dry bulk density (g cm$^{-3}$) | $^{210}Pb$ activity[a] (Bq kg$^{-1}$) | $^{226}Ra$ activity (Bq kg$^{-1}$) | $^{137}Cs$ activity (Bq kg$^{-1}$) | TOC (%) | P (%) |
|---|---|---|---|---|---|---|---|
| GQ-1 | 1 | 0.99 | 181.9 ± 25.7 | 0.7 ± 0.2 0.2 | BDL[b] | 5.35 | 7.43 |
| GQ-2 | 2 | 1.08 | 63.9 ± 20.0 | BDL | BDL | /[c] | / |
| GQ-3 | 3 | 0.95 | 47.3 ± 5.4 | 10.0 ± 1.2 | 2.3 ± 0.6 | 5.29 | 7.18 |
| GQ-4 | 4 | 1.01 | 19.4 ± 3.7 | BDL | 1.6 ± 0.7 | / | / |
| GQ-5 | 5 | 1.06 | 15.8 ± 7.9 | BDL | BDL | 4.32 | 8.05 |
| GQ-6 | 6 | 1.07 | 19.5 ± 3.7 | 8.4 ± 1.3 | BDL | / | / |
| GQ-7 | 7 | 1.12 | 10.7 ± 1.5 | 10.4 ± 1.1 | BDL | 3.29 | 7.94 |
| GJ3-1 + 2 | 1.5 | 1.18 | 60.6 ± 8.9 | 22.1 ± 2.4 | BDL | 1.40 | 0.85 |
| GJ3-3 | 3 | 1.21 | 43.0 ± 7.6 | 8.4 ± 1.2 | 0.8 ± 0.5 | 1.49 | 0.97 |
| GJ3-4 | 4 | 1.18 | 49.2 ± 12.5 | 3.1 ± 0.8 | 0.8 ± 0.8 | 2.22 | 1.12 |
| GJ3-5 | 5 | 1.14 | 54.6 ± 10.2 | 7.1 ± 1.2 | 1.3 ± 0.7 | 2.77 | 1.10 |
| GJ3-6 | 6 | 1.13 | 32.4 ± 8.1 | 1.0 ± 0.3 | 3.1 ± 0.8 | 3.00 | 1.15 |
| GJ3-7 | 7 | 1.13 | 44.9 ± 9.7 | 9.7 ± 1.7 | 0.9 ± 0.7 | 2.51 | 1.09 |
| GJ3-8 | 8 | 1.16 | 33.0 ± 10.9 | 10.5 ± 2.0 | 1.7 ± 0.8 | 2.24 | 1.24 |
| GJ3-9 | 9 | 1.17 | 18.4 ± 5.7 | 15.1 ± 2.2 | 1.1 ± 0.7 | 1.73 | 1.14 |
| JQ1-1 | 1 | 0.98 | 81.4 ± 13.7 | 20.8 ± 2.7 | 2.3 ± 0.9 | 4.72 | 2.08 |
| JQ1-2 | 2 | 0.93 | 56.8 ± 8.7 | 21.9 ± 2.1 | 3.6 ± 0.8 | 5.49 | / |
| JQ1-3 | 3 | 0.96 | 85.4 ± 14.5 | 16.8 ± 2.8 | 1.3 ± 1.0 | 4.50 | 2.18 |
| JQ1-4 | 4 | 0.94 | 94.9 ± 16.1 | 29.3 ± 3.3 | 2.4 ± 1.1 | 5.27 | / |
| JQ1-5 | 5 | 0.89 | 155.3 ± 19.3 | 14.3 ± 2.4 | 3.8 ± 1.1 | 6.46 | 1.76 |
| JQ1-6 | 6 | 0.95 | 88.3 ± 15.1 | 22.2 ± 2.8 | 3.6 ± 0.9 | 4.06 | / |
| JQ1-7 | 7 | 1.01 | 32.0 ± 5.8 | 14.6 ± 1.6 | 3.0 ± 0.6 | 3.04 | / |
| JQ1-8 | 8 | 1.02 | 35.7 ± 9.2 | 14.4 ± 2.2 | 2.1 ± 0.9 | 2.85 | / |
| JQ1-9 | 9 | 1.01 | 15.2 ± 4.8 | 13.7 ± 2.1 | BDL | 2.23 | 2.12 |
| JY2-1 | 1 | 1.00 | 52.1 ± 7.7 | 12.3 ± 1.1 | BDL | 3.86 | 1.83 |
| JY2-2 | 2 | 1.05 | 89.2 ± 11 | 0.88 ± 0.2 | 2.7 ± 0.8 | 3.86 | 1.71 |
| JY2-3 | 3 | 1.04 | 71.0 ± 10 | 5.9 ± 1 | 2.1 ± 0.7 | 4.57 | 1.86 |
| JY2-4 | 4 | 1.04 | 102.1 ± 17.2 | 4.3 ± 0.9 | 2.2 ± 1.1 | 4.02 | 1.73 |
| JY2-5 | 5 | 1.06 | 47.8 ± 9.1 | 8.2 ± 1.2 | BDL | 4.25 | 1.66 |
| JY2-6 | 6 | 1.06 | 40.2 ± 7.1 | 11.3 ± 1.3 | 2.0 ± 0.7 | 3.36 | 1.73 |
| JY2-7 | 7 | 1.08 | 45.9 ± 9 | 13.1 ± 2 | 2.4 ± 1.0 | 3.89 | 2.13 |
| JY2-8 | 8 | 1.08 | 25.3 ± 4.8 | 1.1 ± 0.2 | 3.0 ± 0.7 | 2.71 | 1.71 |
| JY2-9 | 9 | 1.14 | 21.7 ± 4.2 | 9.6 ± 1 | 2.4 ± 0.8 | 2.53 | 1.71 |
| JY2-10 | 10 | 1.13 | 8.6 ± 1.7 | 4.1 ± 0.6 | 1.4 ± 0.5 | 2.31 | 1.77 |
| JY2-11 | 11 | 1.11 | 10.3 ± 3 | 7.5 ± 1.1 | BDL | / | / |

*Note* [a]The unit of all the activity values is in Bq kg$^{-1}$ dry weight. [b]*BDL* below detection limit. [c]'/' means unmeasured. Reprinted from Xu et al. (2010). Copyright (2010), with permission from Elsevier

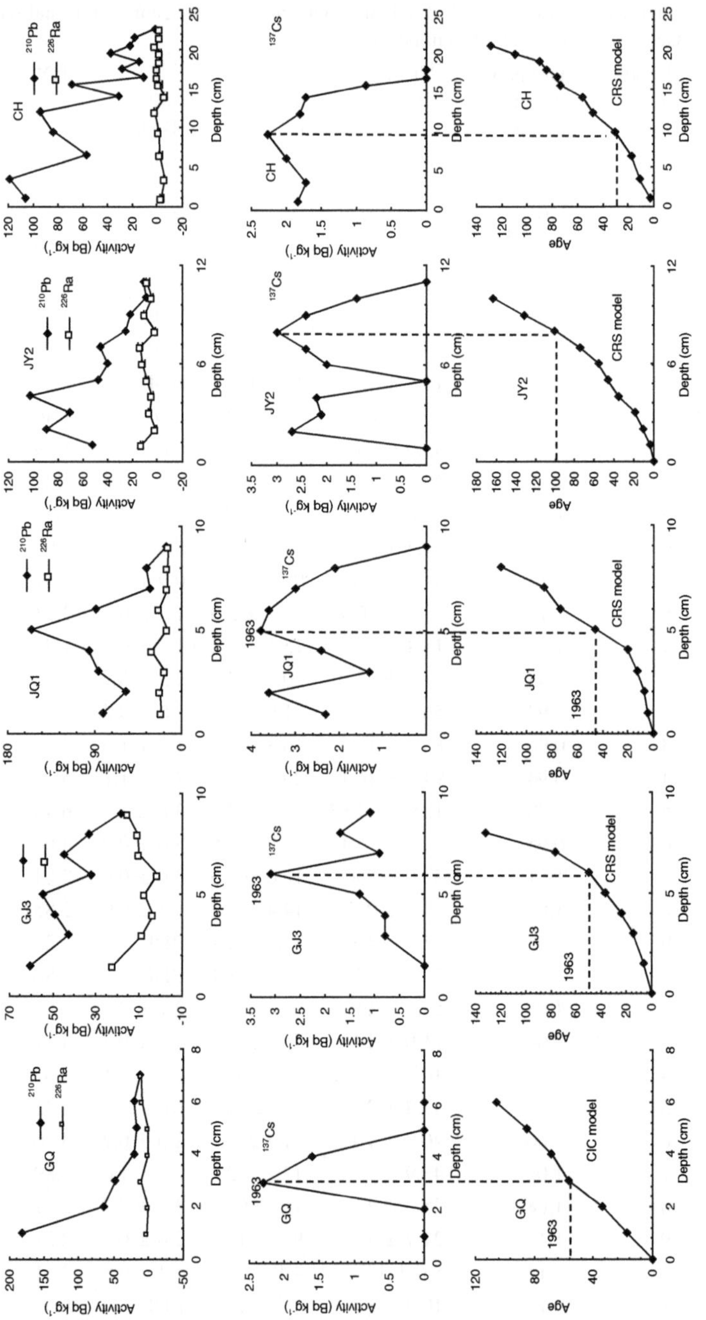


**Fig. 4.2** $^{210}$Pb chronology of GQ, GJ3, JQ1, JY2 and CH

(F), in which $\lambda = 0.0311\ a^{-1}$, the decay constant of $^{210}$Pb. Based on this equation, the calculated excess $^{210}$Pb supply rates for GQ, JQ1, GJ3, and JY2 are $113 \pm 15$ Bq m$^{-2}$ year$^{-1}$, $139 \pm 16$ Bq m$^{-2}$ year$^{-1}$, $108 \pm 12$ Bq m$^{-2}$ year$^{-1}$ and $143 \pm 16$ Bq m$^{-2}$ year$^{-1}$, respectively. The average $^{210}$Pb supply rate of these different islands is $126 \pm 15$ Bq m$^{-2}$ year$^{-1}$, almost completely equivalent to the mean flux of northern hemisphere of 125 Bq m$^{-2}$ a$^{-1}$ (Xiang 2000); and the supply rate of excess $^{210}$Pb in the Xisha Islands shows evident hemispherical distribution features and did not show significant notable regional differences.

### *4.1.4 Potential Impacts of Anthropogenic Nuclear Test on the Xisha Islands*

Artificial radionuclides, e.g. $^{137}$Cs and $^{241}$Am, arise from human nuclear test and the occurrence of these anthropogenic isotopes is an ideal marker for human nuclear activities. As shown in Table 4.1, the $^{137}$Cs activity concentrations in the bulk sediments are pretty low (less than 5 Bq kg$^{-1}$) in the Xisha islands. Of the five sediment profiles, the maximum inventory of $^{137}$Cs was 210 Bq m$^{-2}$ for JQ1. Compared with the mean values of about 1,500 Bq m$^{-2}$ in the northern hemisphere and 500 Bq m$^{-2}$ in the southern hemisphere (Appleby et al. 1995), the cumulative $^{137}$Cs fallout in the Xisha Islands is mucher lower, suggesting a minor human impact. In addition to the radionuclide $^{137}$Cs, we also examined another artificial $^{241}$Am, and the activities of this radioisotope in the five sediment cores were under the detection limit. The extremely low lvels of $^{137}$Cs and $^{241}$Am in the ornithogenic sediments of the Xisha Islands reveal that the accumulation of artificial radionuclides fallout and the impact of human nuclear activities on the Xisha islands are insignificant. This finding is in concert with the findings by Xiang (2000), which suggested that the $^{137}$Cs inventory in the middle-latitude zone reached the maximum level, but it was relatively low in equatorial areas.

According to $^{137}$Cs activity versus depth profiles of JQ1 and JY2 (Fig. 4.2), there is an evident younger peak besides the fallout maximum in 1963. Although we are not sure about the reason for its appearance, Chernobyl nuclear accident in 1986 might be one plausible explanation.

### *4.1.5 Influence of Seabird Activities on Radionuclide Distribution*

Seabirds may significantly impacted on radionuclide concentrations in their immediate surroundings (Dowdall et al. 2003, 2005b). For example, the soils influenced by bird communities exhibited enrichment factors of 5 for $^{238}$U, 8 for $^{137}$Cs and 2 for $^{226}$Ra compared to the nuclide contentrations in soils from control area (Dowdall et al. 2005b). Nevertheless, it remains a question that whether

**Table 4.2** Activities of $^{210}Pb$, $^{226}Ra$ and $^{137}Cs$ in the pure guano of Jinqing Island and Ganquan Island

| Sample No. | Depth (cm) | $^{210}Pb$ activity ($Bq\ kg^{-1}$) | $^{226}Ra$ activity ($Bq\ kg^{-1}$) | $^{210}Pb_{ex}$ activity ($Bq\ kg^{-1}$) | $^{137}Cs$ activity ($Bq\ kg^{-1}$) |
|---|---|---|---|---|---|
| JQ-1 | 1 | 87.1 ± 15.9 | 48.7 ± 4.5 | 38.4 ± 7.9 | BDL |
| JQ-2 | 2 | 63.5 ± 12.4 | 15.8 ± 2.2 | 47.7 ± 11.5 | |
| JQ-4 | 4 | 75.7 ± 17.7 | 48.5 ± 5.8 | 27.1 ± 7.1 | |
| JQ-5 | 5 | 36.6 ± 7.8 | 26.1 ± 2.7 | 10.4 ± 2.5 | |
| JQ-6 | 6 | 10.5 ± 4.1 | 4.4 ± 1.1 | 6.2 ± 2.9 | |
| JQ-7 | 7 | 24.2 ± 6.2 | 22.0 ± 3.3 | 2.2 ± 0.7 | |
| GQ1-2 | 3 | 0.68 ± 0.04 | BDL[a] | 0.68 ± 0.04 | BDL |
| GQ1-3 | 5 | 8.17 ± 0.52 | BDL | 8.17 ± 0.52 | |
| GQ1-4 | 7 | 6.84 ± 0.44 | 6.12 ± 0.33 | 0.72 ± 0.06 | |

*Note* [a]*BDL* below detection limit. Reprinted from Xu et al. (2010). Copyright (2010), with permission from Elsevier

the radionuclide enrichment is directly linked with seabird droppings or indirectly caused by the increase in nutrient-rich materials, e.g. guano samples. We also tested the activity concentrations of $^{210}Pb$ and $^{226}Ra$ in the pure guano particles separated from the bulk ornithogenic sediments, and the results are given in Table 4.2 and Fig. 4.3. As shown in Fig. 4.3, the excess $^{210}Pb$ concentrations in both GQ1 and JQ cores are lower than that of bulk sediments in the same sediment layer. Consequently, guano is not the principal source of $^{210}Pb$ in the bulk sediments. It is also concluded that though seabirds are top predators in marine food webs, $^{210}Pb$ is not biomagnified.

In the Xisha Islands, a large number of bird droppings, i.e. guano pellets, have been accumulated in the widely distributed ornithogenic sediments. The nutrient element phosphorus is significantly enriched in guano particles, and it acts as a major nutrient source for the island ecosystems. In general, the levels of phosphorus increase with guano content. We performed the Pearson correlation analysis on

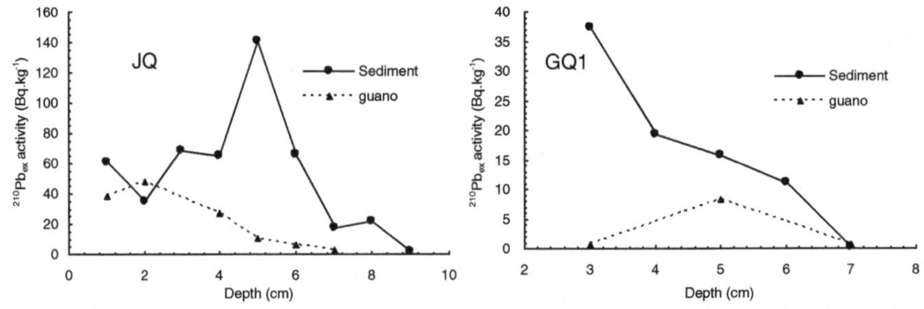


**Fig. 4.3** Comparison of $^{210}Pb_{ex}$ activities between the ornithogenic coral sand sediments and the guano separated from bulk sediments. Reprinted from Xu et al. (2010). Copyright (2010), with permission from Elsevier

the excess $^{210}Pb$ and phosphorus concentration, and an insignificant correlation was obtained (Fig. 4.4a, R = 0.13), unveiling that the excess $^{210}Pb$ in the sediments is not derived from bird droppings. Furthermore, the $^{137}Cs$ activities in all guano samples were under the detection limit of gamma spectrometry, and thus bird-derived guano was not the source of $^{137}Cs$ either. Michelutti et al. (2008) examined the possible impacts of seabird activity on the $^{210}Pb$ deposition at Cape Vera of Devon Island in Arctic ponds, a large seabird colony of northern fulmars. They found that the fluxes of $^{210}Pb$ were similar to, or less than, those measured at other Arctic locations not influenced by seabirds. This is in consistent with our finding.

The increase in excess $^{210}Pb$ concentrations in the sediments is likely related to the elevated levels of nutrient-rich materials. As illustrated in Fig. 4.4b, excess $^{210}Pb$ activities and TOC in the ornithogenic sediments are significantly correlated (R = 0.63). Some radionuclides have a high affinity with organic matter, and organics could adsorb radionuclides and thus further enhance the excess $^{210}Pb$ activities in soils or sediments (Dowdall et al. 2005a; Nozaki et al. 1978). Organic matter may also increase soil or sediment moisture and help in retaining radionuclides. Since nutrient-rich seabird guano increases organic matter production in the bulk sediments, seabirds can indirectly lead to the increase in $^{210}Pb$ concentrations. Plant humus may also play a similar role.

Hence, the excess $^{210}Pb$ and $^{137}Cs$ in the ornithogenic coral sand sediments are mainly derived from atmospheric precipitation, rather than seabird guano. Although organic matter is not a significant source, the organics derived from guano and plant humus could further enhance radionuclides.

Based upon the activity concentrations of several radionuclides in the ornithogenic coral sand sediments and the separated guano particles from the Xisha Islands, it is concluded that:

1. $^{210}Pb/^{226}Ra$ equilibrium is reached at a certain depth of all the five sediment profiles containing guano and coral sands. The $^{210}Pb$ dating technique

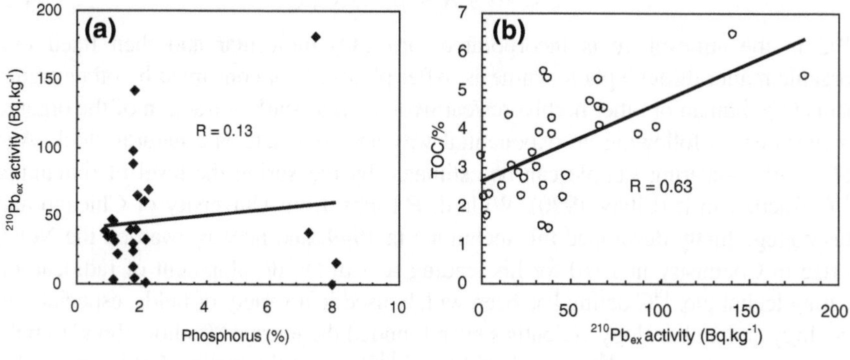


**Fig. 4.4** Correlations between $^{210}Pb_{ex}$ activity and phosphorus concentration (**a**), TOC content (**b**) in the ornithogenic coral sand sediments on the Xisha Islands. Reprinted from Xu et al. (2010). Copyright (2010), with permission from Elsevier

is applicable for the widely distributed ornithogenic sediments in the Xisha Islands, and this builds up the basis for our further palaeoecology studies.

2. The mean atmospheric $^{210}Pb$ flux in the Xisha Islands is very close to the average $^{210}Pb$ flux of the northern hemisphere. Human nuclear tests did not significantly impact on radionuclide fallout on the Xisha Islands.
3. Atmospheric precipitation is the most important source for radionuclides in the ornithogenic coral sand deposits, and the organic matter from plant debris and guano could further enhance them.

## 4.2 Radiocarbon Dating and Established Chronology of the Studied Sediment Cores

In addition to $^{210}Pb$ dating, radiocarbon analysis constructs the fundamental part and is essential for the following part of this thesis. A brief introduction to radiocarbon ($^{14}C$), including its principles and applications, is given below.

### *4.2.1 About $^{14}C$ Dating*

1. Basic physics

The element carbon has three isotopes, including radioactive $^{14}C$ and stables $^{12}C$ and $^{13}C$. Carbon-14 is a naturally occurring and radiometric isotope of carbon. It comes from nitrogen-14 (formula 4.1) and then spontaneously decays to nitrogen-14 (formula 4.2) in the atmosphere.

$$ {}_{0}^{1}n + {}_{7}^{14}N \rightarrow {}_{6}^{14}C + {}_{1}^{1}H \tag{4.1}$$

$$ {}_{6}^{14}C \rightarrow {}_{7}^{14}N + \beta^{-} + \overline{\nu} + Q \tag{4.2}$$

$^{14}C$ in the atmosphere is incorporated into $CO_2$ molecular and then fixed into organic matter through photosynthesis. After plants die or consumed by other organisms (e.g. human or other herbivores/carnivores), radiocarbon fraction of the organic matter decays following an exponential way and thus acts as a natural clock. Age of a carbon-bearing sample can be estimated by measuring the level of remaining $^{14}C$ fraction in it (Libby 1970). Willard. F Libby from University of Chicago and his college firstly developed this technique in 1949, and he was awarded the Nobel Prize in Chemistry in 1970 for his leading role in the development of radiocarbon dating technique. $^{14}C$ dating has been widely used in a variety of fields, especially in geology and archaeology. Scientists even founded the journal "Radiocarbon" to publish papers relevant to $^{14}C$. The half-life of $^{14}C$ was estimated as 5,568 years (also know as Libby half-life), but corrected to a more accurate value 5,730 years at a conference in 1962. It should be noted that, although Libby half-life (5,568 years) is not accurate, it continues to be used nowadays. Reports of $^{14}C$ ages from different labs should be accompanied by the half-life, as different half-life values can impact $^{14}C$ age of a sample to some extent. There are two analytical techniques to detect radioactivity of $^{14}C$: Accelerated Mass Spectrometry (AMS) and liquid scintillation counter. Of these two techniques, the precision of liquid scintillation counter is relatively low, while AMS analysis is costly and only national laboratories or equivalent institutions can offer such equipments. Recently, Galli et al. (2011) developed a new technique to analyze ultratrace $^{14}C$ by spectrum. Once this approach is successfully commercialized, radiocarbon dating will be more widely used (Zare 2012).

2. Calibration

One theoretical assumption of radiocarbon dating is that atmospheric $^{14}C$ remains constant over the past tens of thousands of years. In fact, the level of $^{14}C$ in the atmosphere has not been strictly constant in the past, due to changes in the cosmic ray intensity. Thus, the determined raw $^{14}C$ age (or conventional age) needs to be calibrated (Cal age). Since radiocarbon is not the major point of the present study, details of calibration are referred to several papers concerning $^{14}C$ age calibration shown below. So far, there have been three widely-cited calibration datasets from the same research group (M Stuiver and his colleagues) and all of them were published by the scientific journal "Radiocarbon". The datasets are INTCAL98 (Stuiver et al. 1998), INTCAL04 (Reimer et al. 2004) and INTCAL09 (Reimer et al. 2009), whose calibration range are 0–24, 0–26 and 0–50 ka BP, respectively. The above-mentioned datasets and related materials can be downloaded from webpage of Quaternary Isotope Laboratory, the University of Washington (http://depts.washington.edu/qil/). Furthermore, according to published calibration data, the calibration curves are not monotonous, implying that one conventional $^{14}C$ age may corresponds to two, or more, calibration ages. Under such circumstances, additional information, e.g. geological settings, has to be considered to confirm the date.

### 4.2.2 *Radiocarbon Analysis of Bird Bone Samples from the Xisha Islands*

A large number of well-preserved carbon-bearing biololgical remains, including bird and fish bones, were observed from the collected ornithogenic sediments. Collagen is naturally occurring protein and can be well-preserved for quite a long time. Bone collagen is thus ideal material for radiocarbon analysis. AMS$^{14}C$ analyses of several bird bone samples were performed at the University of California, Irvine, or Beta Analytic, Inc., Florida. The AMS$^{14}C$ ages of bone samples were calibrated into calendar year before present (Cal BP), using the internationally recommended INTCAL98 calibration dataset proposed by Stuiver et al. (1998) and the CALIB 4.3 program (Stuiver and Reimer 1993). When calibrating, ΔR (the difference between the regional and global marine $^{14}C$ ages) was assumed as $-25 \pm 20$ year, and the marine carbon component as 100 %. The results of AMS$^{14}$C analysis of bone samples from GQ, GJ3, JQ and CH profiles are supplied in Table 4.3. $^{14}$C ages of JY2 is not included in this table, as we failed to collect sufficient avian bone samples for radiocarbon analysis from this sediment core.

### 4.2.3 Establishment of Chronology for the Studied Cores GQ, GJ3, JQ, CH and JY2

Joint use of $^{210}$Pb ages and $^{14}$C dates can supply more accurate chronology (Cooke et al. 2010). The chronology of these five cores is thus established by these two dating techniques. The combined chronology, i.e. age-depth model, of the studied cores GQ, GJ3, JQ, CH and JY2 are supplied in Fig. 4.5.

From Fig. 4.5, the sediment core GQ has the longest history, dating back to approximately 2,200 years. In contrast, other cores contain sediments within approximately the last millennium. Additionally, according to the chronology

**Table 4.3** $^{14}$C ages of bone samples in the studied sediment cores

| Core No | Sample No | Material | Depth (cm) | Conventional age BP | Calibrated age BP (2σ) |
|---|---|---|---|---|---|
| GQ | GQ1-6 | Bone | 11 | 560 ± 20 | 256 (283–142) |
| | GQ1-10 | Bone | 19 | 640 ± 15 | 295 (334–270) |
| | GQ1-25 | Bone | 49 | 1,845 ± 15 | 1,403 (1,485–1,340) |
| | GQ1-31 | Bone | 61 | 2,150 ± 20 | 1,773 (1,836–1,696) |
| | GQ1-37 | Bone | 73 | 2,355 ± 15 | 1,992 (2,066–1,933) |
| | GQ1-43 | Bone | 85 | 2,450 ± 20 | 2,117 (2,183–2,042) |
| | GQ1-46 | Bone | 100 | 2,495 ± 20 | 2,158 (2,282–2,106) |
| GJ3 | GJ3-21 | Bone | 27.3 | 625 ± 20 | 287 (322–260) |
| | GJ3-32 | Bone | 41.6 | 770 ± 20 | 447 (488–402) |
| | GJ3-43 | Bone | 55.9 | 860 ± 20 | 502 (533–466) |
| | GJ3-55 | Bone | 71.5 | 1,005 ± 20 | 619 (649–545) |
| | GJ3-60 | Bone | 78 | 1,115 ± 15 | 673 (723–649) |
| | GJ3-68 | Bone | 88.4 | 1,085 ± 20 | 658 (705–631) |
| | GJ3-75 | Bone | 97.5 | 1,150 ± 15 | 705 (759–662) |
| | GJ3-81 | Bone | 105 | 1,175 ± 20 | 728 (786–671) |
| JQ1 | JQ1-33 | Bone | 44 | 1,175 ± 15 | 728 (783–674) |
| | JQ1-38 | Bone | 51 | 1,255 ± 20 | 824 (901–745) |
| CH | CH-26 | Bone | 37 | 620 ± 30 | 284 (338–254) |
| | CH-38 | Bone | 55 | 740 ± 20 | 424 (468–323) |
| | CH-50 | Bone | 72 | 1,260 ± 30 | 829 (912–736) |

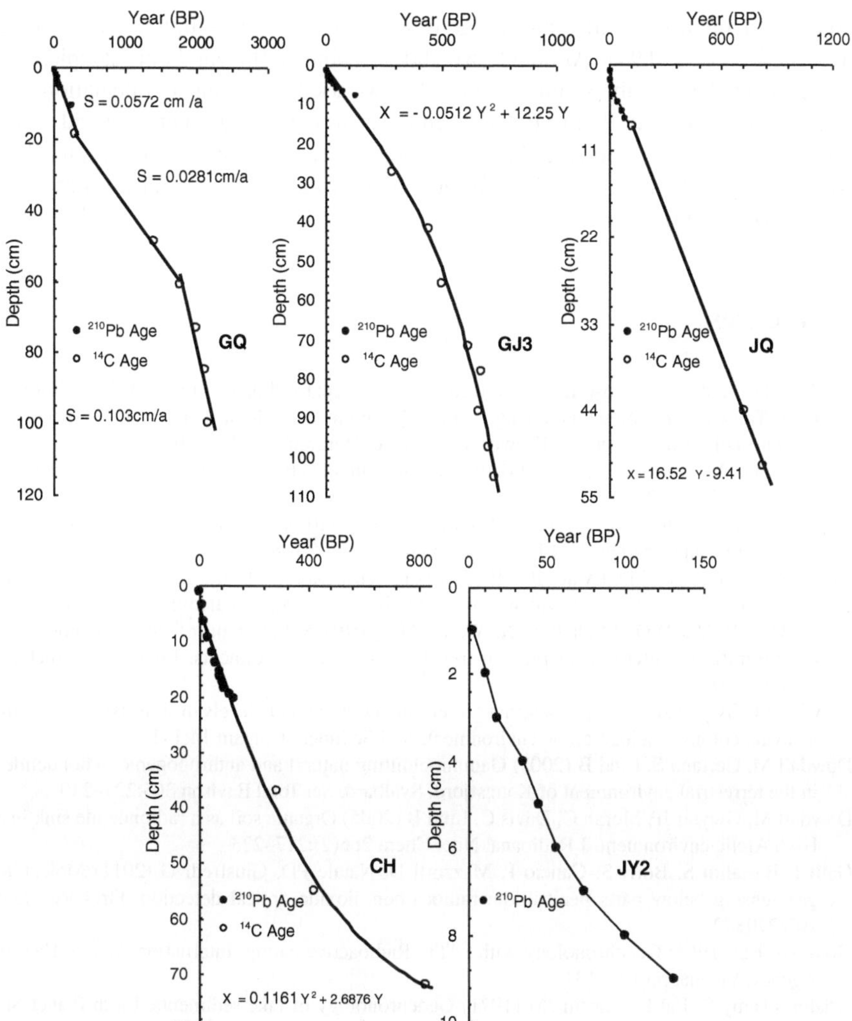


**Fig. 4.5** Chronology of profiles GQ, GJ3, JQ, CH and JY2

of GQ, the ages do not show a linear trend for the whole profile and the inferred sedimentation rates are divided into three stages. Mean sedimentation rates of interlayer 0–20, 20–60 and 60–107 cm are 0.0572, 0.0281 and 0.103 cm year$^{-1}$, respectively. An excellent quadratic interpolation between $^{210}$Pb and $^{14}$C dates was applied to assign ages down the cores CH and GJ3, and the age-depth models for CH and GJ3 can be described by the following equations: $X = 0.12\ Y^2 + 2.69\ Y$ and $X = -0.051\ Y^2 + 12.25\ Y$, respectively, in which X and Y are the age and depth of the cores, respectively. For the profile JQ, however, a good linear regression equation was used to calculate the dates below 8 cm depth (Fig. 4.5).

Age-depth relation of the profile JQ is also shown in this figure (X = 16.52 Y − 9.41, R = 0.99). Although we did not find sufficient bone samples for AMS$^{14}$C analysis in the sediment core JY2, we used the mean sedimentation rate in the layer from 3 to 9 cm i.e. 0.073 cm/year, to extrapolate dates beyond $^{210}$Pb sediments. This is acceptable due to the homogeneous lithology and a quite low water content of the profile. The chronologies of the five sediment profiles were summarized in Fig. 4.5.

## References

Appleby PG (2001) Chronostratigraphic techniques in recent sediments. In: Last WM, Smol JP (eds) Tracking environmental change using lake sediments volume 1: basin analysis, coring and chronological techniques. Kluwer Academmic, Dordrech, pp 171–203

Appleby PG (2008) Three decades of dating recent sediments by fallout radionuclides: a review. Holocene 18:83–93

Appleby PG, Oldfield F (1978) The calculation of lead-210 dates assuming a constant rate of supply of unsupported lead-210 to the sediment. Catena 5:1–8

Apple PG, Jones VJ, Ellis-Evans JC (1995) Radiometric dating of lake sediments from Signy Island (maritime Antarctic): evidence of recent climatic change. J Paleolimnol 13:179–191

Cooke CA, Hobbs WO, Michelutti N, Wolfe AP (2010) Reliance on $^{210}$Pb chronology can compromise the inference of preindustrial Hg flux to lake sediments. Environ Sci Technol 44:1998–2003

Dowdall M, Gwynn JP (2005) Assessment of elevated radionuclide levels in soils associated with an avian colony in a high arctic environment. Soil Sediment Contam 14:1–11

Dowdall M, Gerland S, Lind B (2003) Gamma-emitting natural and anthropogenic radionuclides in the terrestrial environment of Kongsfjord, Svalbard. Sci Total Environ 305:229–240

Dowdall M, Gwynn JP, Moran C, Davis C, Lind B (2005) Organic soil as a radionuclide sink in a High Arctic environment. J Radioanal Nucl Chem 266(2):217–223

Galli I, Bartalini S, Borri S, Cancio P, Mazzotti D, Natale PD, Giusfredi G (2011) Molecular gas sensing below parts per trillion: radiocarbon-dioxide optical detection. Phys Rev Lett 107:270802

Goldberg ED (1963) Geochronology with $^{210}$Pb. Radioactive dating. Internation Atomic Energy Agency, Vienna, pp 121–131

Krishnaswamy S, Lal D, Martin JM (1971) Geochronology of lake sediments. Earth Planet Sci Lett 11(5):407–414

Libby WF (1970) Radiocarbon dating. Philos Trans R Soc A Math Phys Eng Sci 269:1–10

Michelutti N, Blais JM, Liu H, Keatley BE, Douglas MSV, Mallory ML, Smol JP (2008) A test of the possible influence of seabird activity on the $^{210}$Pb flux in high arctic ponds at Cape Vera, Devon Island, Nunavut: implications for radiochronology. J Paleolimnol 40:783–791

Nozaki Y, Demaster DJ, Lewis DM, Turekian KK (1978) Atmospheric $^{210}$Pb flux determined from soil profiles. J Geophys Res 83:4047–4051

Reimer PJ, Baillie MGL, Bard E, Bayliss A, Beck JW, Bertrand CJH, Blackwell PG, Buck CE, Burr GS, Cutler KB, Damon PE, Edwards RL, Fairbanks RG, Friedrich M, Guilderson TP, Hogg AG, Hughen KA, Kromer B, McCormac G, Manning S, Ramsey CB, Reimer RW, Remmele S, Southon JR, Stuiver M, Talamo S, Taylor FW, Plicht JVD, Weyhenmeyer CE (2004) INTCAL04 terrestrial radiocarbon age calibration, 0–26 cal kyear BP. Radiocarbon 46(3):1029–1058

Reimer PJ, Baillie MGL, Bard E, Bayliss A, Beck JW, Blackwell PG, Ramsey CB, Buck CE, Burr GS, Edwards RL, Friedrich M, Grootes PM, Guilderson TP, Hajdas I, Heaton TJ, Hogg AG, Hughen KA, Kaiser KF, Kromer B, McCormac FG, Manning SW, Reimer RW, Richards

DA, Southon JR, Talamo S, Turney CSM, Plicht JVD, Weyhenmeyer CE (2009) INTCAL09 and marine09 radiocarbon age calibration curves, 0–50000 years cal BP. Radiocarbon 51(4):1111–1150

Stuiver M, Reimer PJ (1993) Extended C-14 data-base and revised calib 3.0 C-14 age calibration programe. Radiocarbon 35:215–230

Stuiver M, Reimer PJ, Bard E, Beck JW, Burr GS, Hughen KA, Kromer B, McCORMAC G, Plicht JVD, Spurk M (1998) INTCAL98 radiocarbon age calibration, 24000-0 cal BP. Radiocarbon 40(3):1041–1083

Wan GJ (1997) $^{210}$Pb dating for recent sedimentation. Quat Sci 3:230–239 (in Chinese with English abstract)

Xiang L (2000) The study of lake environmental change using radionuclides as tracers. PhD dissertation, Nanjing Institute of Geography and Limnology

Xu LQ, Liu XD, Sun LG, Yan H, Liu Y, Luo YH, Huang J, Wang YH (2010) Distribution of radionuclides in the guano sediments of Xisha Islands, South China Sea and its implication. J Environ Radioact 101:362–368

Zhang XB (2005) Discussion on interpretations of $^{137}$Cs depth distribution profiles of lake deposits. J Math Sci 23(3):294–299 (in Chinese with English abstract)

Zare RN (2012) Ultrasensitive radiocarbon detection. Nature 482:312–313

[illegible]

# Chapter 5
# Geochemical Evidence for the Development of Coral Island Ecosystem on the Xisha Archipelago of the South China Sea

Due to geographical and political reasons, the Xisha archipelago has been well-preserved and thus is an ideal place to study developmental history of tropical coral island ecosystem in the ocean.

Both our laboratory lithological studies and field observations show that the widely distributed ornithogenic sediments in the Xisha Islands are mainly composed of plant residue, guano and coral sands. Of these three major constituents, guano and plant humus determine the input of organic matter (OM), while coral sand constitutes the source of inorganic materials. Changes in source materials are essential for examining the evolution of such island ecosystems. Developmental modes of these small coral islands are investigated by geochemical analysis of the seabird-influenced sediments in this chapter.

## 5.1 Source Materials of the Ornithogenic Sediments

The vertical distributions of total nitrogen (TN), total organic carbon (TOC), total hydrogen (TH) and calcium carbonate ($CaCO_3$) for profiles GQ, GJ3, JQ and JY2 are given in Fig. 5.1. Despite the evident fluctuations, TOC, TN and TH in each sediment profile yielded quite similar distribution patterns. It was also noted that levels of TOC, TN and TH in the upper layers were much higher than those below the depths 11 cm (GQ), 9 cm (JQ), 16 cm (GJ3) and 12 cm (JY2), which is in accordance with lithological observations. A Pearson correlation test between TOC and TN in the four profiles was performed, and significant positive correlations were obtained in all the sediment profiles (Fig. 5.2, $R > 0.97$, $p < 0.01$). The intercepts of the regression equation on the Y axis, indicative of the level of inorganic nitrogen in the sediments, were less than 0.04 %, and were much smaller than the TN concentrations (Fig. 5.1). Consequently, TN in the ornithogenic sediments is mainly derived from OM; TOC, TN and TH can represent relative levels of OM.



L. Xu, *Impact of Climate Change and Human Activity on the Eco-environment*, Springer Theses, DOI 10.1007/978-3-662-45003-1_5

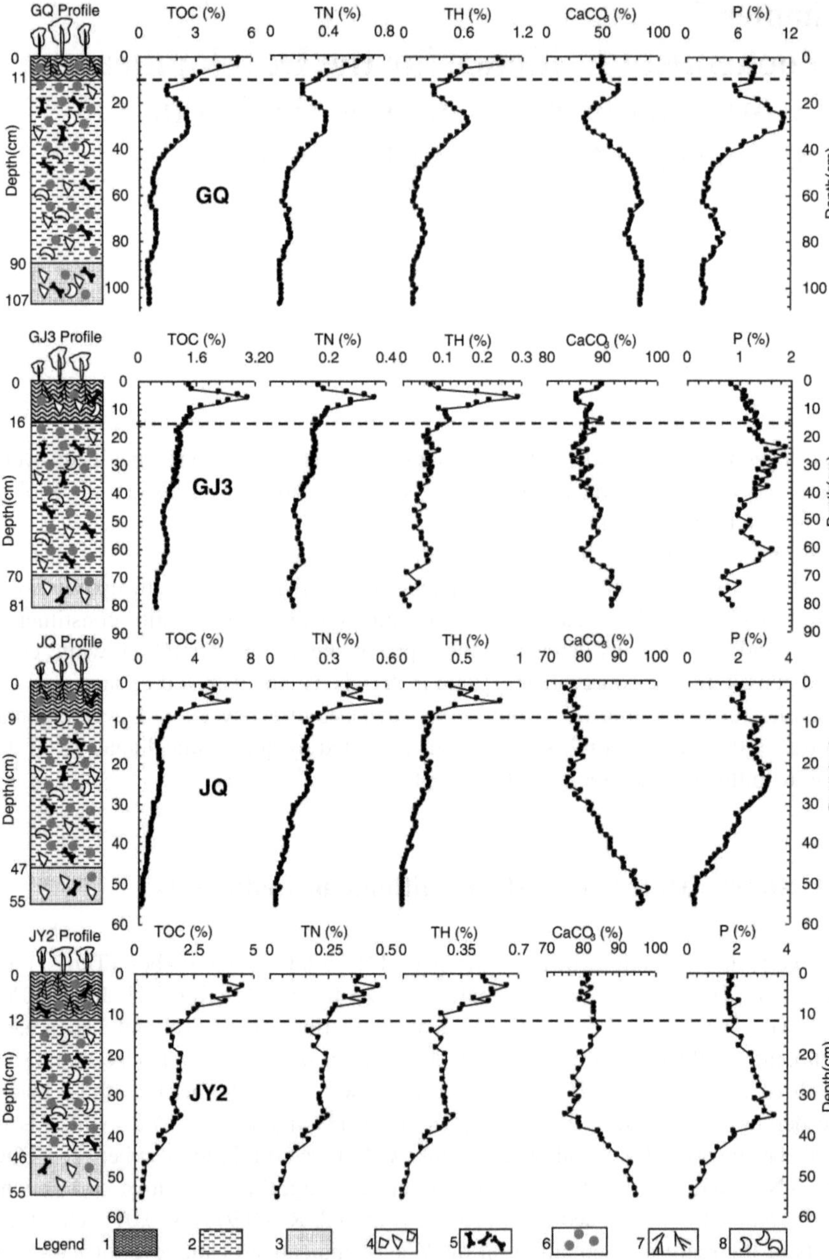


**Fig. 5.1** Vertical distributions of TOC, TN, TH, $CaCO_3$ and P for the profiles GQ, GJ3, JQ and JY2. Legend: *1 grey black* humus, plant residues and medium to large size coral sands in the upper layer, a little bit guano and seabird bone remains; *2 light brown* ornithogenic sediments, containing a lot of medium to large size coral sands, guano particles, bone remains and fish scales; *3 yellow to white* coral sand sediment layer with low organic matter content, a few guano pellets and numerous calcareous bioclasts; *4* calcareous bioclasts; *5* bird and fish bones; *6* guano; *7* remains of plant leaf, stem and root; *8* fish scales (reprinted from Xu et al. (2011b), Copyright (2011), with permission from Elsevier)

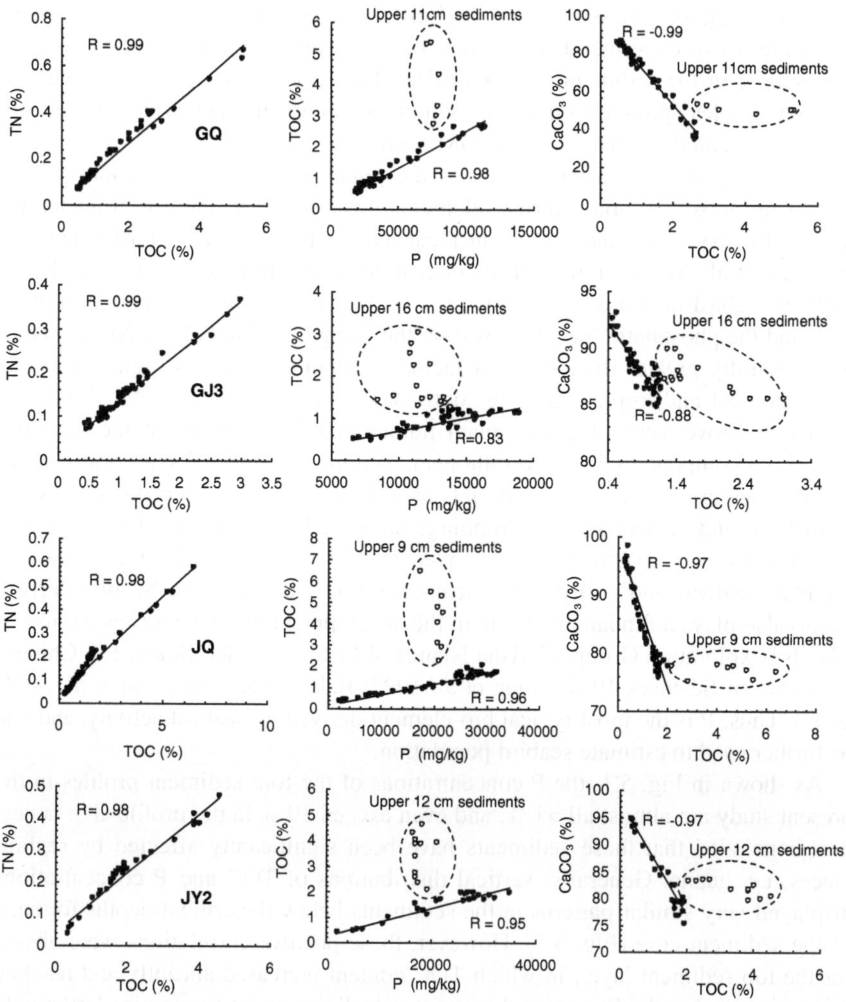


**Fig. 5.2** Correlations between TN, TOC, P and $CaCO_3$ in the sediment profiles GQ, GJ3, JQ and JY2. Legend: O samples in the sediments above 11, 16, 9 and 12 cm depth in GQ, GJ3, JQ and JY2 respectively (reprinted from Xu et al. (2011b), Copyright (2011), with permission from Elsevier)

As shown in Fig. 5.1, TOC concentrations below specific depths of 11 cm (GQ), 9 cm (JQ), 16 cm (GJ3) and 12 cm (JY2) were significantly and negatively correlated with coral sand-derived $CaCO_3$ in the sediments (Fig. 5.2). These specific depths are defined as “critical depths”. In contrast, the OM increased rapidly in the sediments above these critical depths, but $CaCO_3$ remained at a relatively stable level (Fig. 5.1). According to field observations, clastic coral sands are the pedogenic parent materials for ornithogenic sediments in the Xisha archipelago (Liu et al. 2006, 2008), and coral sand is the overwhelming constituent in each of the four sediment core, consistent with the high levels of $CaCO_3$ in all the sediment profiles (Fig. 5.1). The significantly negative relationship between $CaCO_3$

and TOC suggests a possible "dilution" effect of OM input by coral sands, likely reflecting a two-endermember structure of the sediments and a relatively simple OM source in the sediments below critical depths. However, the significant correlation was less pronounced for the sediments above the critical depths and this may be explained by changes in the OM source.

According to our earlier studies of the guano sediments, phosphorus (P) is one of the most essential nutrient elements transferred by seabirds from marine to island ecosystems, and thus is an ideal marker for relative seabird population size (Liu et al. 2006). In general, P concentration in subtropical and tropical soils without seabird influence is less than 0.1 %, but its level in the guano phosphate rock and the phosphatic limestone soil on the Dongdao Island of the Xisha Islands was generally greater than 10 %, a factor of about 5–30 times higher than that in subtropical and tropical soils in southern China (Gong et al. 1996, 1997). As discussed above, seabird guano, plant humus and coral sands are the predominant three components in the ornithogenic sediments of the Xisha coral islands, and P concentrations in those materials have been analyzed in our previous work. Both fresh and ancient seabird droppings have the highest levels of P, as high as ~15 % (Liu et al. 2006). Seabird-derived guano acts as a major nutrient source for plant growth and soil development on tropical coral atolls. Seabird-derived guano also plays a fundamental role in the development of phosphorites on insular islands (Exploration Group of Xisha Islands of Institute of Soil Science of Chinese Academy of Sciences 1977; Gong et al. 1997; Baker et al. 1998; Liu et al. 2006, 2008). Thus, P is the most typical bio-element derived by seabird activity, and can be further used to estimate seabird population.

As shown in Fig. 5.1, the P concentrations of the four sediment profiles in the present study are almost all >1 %, and even exceed 10 % in the profile GQ; hence, it is concluded that these sediments have been significantly affected by seabird faeces, i.e. guano. Generally, vertical distributions of TOC and P concentrations displayed very similar patterns in the sediments below the critical depths for each of the sediment core (Fig. 5.1). However, these positive correlations were absent for the top sediment layer, in which TOC content increased abruptly and reached its maximum, but the P content displayed a declining trend or remained relatively constant at low levels, possibly implying a change of material source in the surface layer. The Pearson correlation analysis showed that P and TOC in the sediments below the critical depths bear a significant and positive correlation (Fig. 5.2), suggesting that they probably shared a common source, i.e. guano. However, as illustrated by Fig. 5.2, the positive correlation between TOC and P was abruptly ended in the upper layer. This result, in combination with the evident lack of negative relationship between $CaCO_3$ and TOC (Fig. 5.2), suggests that the source of organic matter in the bulk sediments has changed above these critical depths.

We also determined concentrations of TOC and TN in the three main sediment components of the Xisha Islands, i.e. plant humus, and guano and coral sands (Fig. 5.3). TOC and TN concentrations were highest in plant humus (as high as 38.05 % and 4.34 %, respectively). Their levels (0.41 % for TOC, $n = 20$, and 0.037 % for TN, $n = 10$) were lowest in coral sands; and guano samples had the intermediate

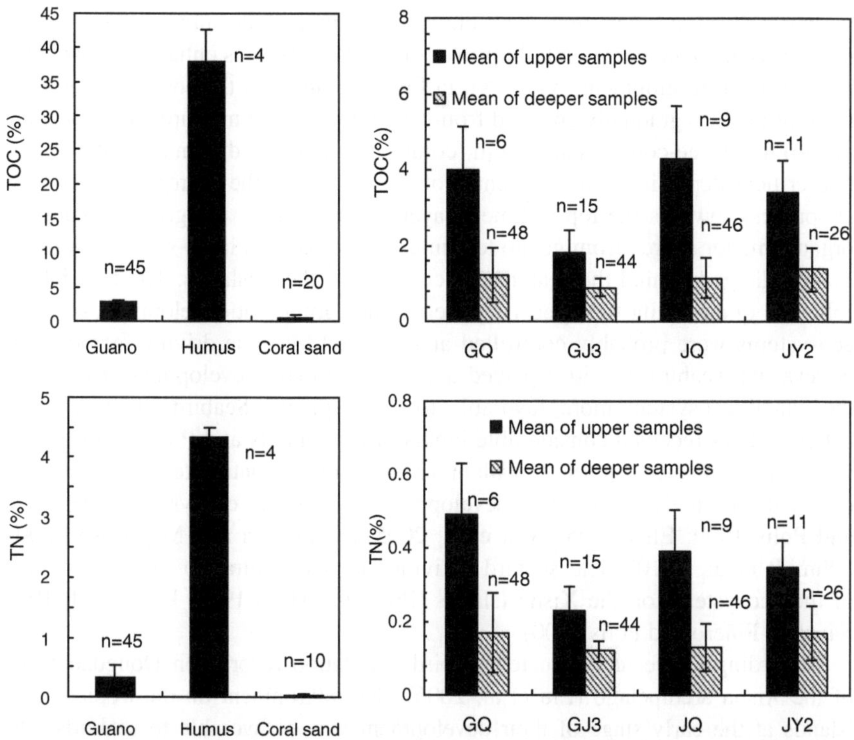


**Fig. 5.3** TOC and TN levels in each sediment profile, and for each source material (Reprinted from Xu et al. (2011b), Copyright (2011), with permission from Elsevier.)

concentrations of TOC and TN (2.82 % for TOC and 0.33 % for TN; n = 45). Furthermore, both TOC and TN contents in the upper layer samples were significantly greater than those in samples below the critical depths (Fig. 5.3). Considering the occurrence of large numbers of guano particles in the sediments below the critical depths of the four cores, we suggest that the organic matter below the critical depths is mainly derived from guano, and that the rapid increase of organic matter in the upper layers of the profiles is largely attributed to the input of plant humus.

Coral sand is the most important soil-forming material on the coral islets of Xisha archipelago, and the pedogenesis on the cay islands is significantly controlled by plant growth and guano input (Wang 2001). As discussed above, both bird droppings and plant humus are sources of the organic matter in four ornithogenic sediment profiles. Based on our earlier study on the Dongdao Island (which is occupied by larger numbers of seabirds in the present day; Liu et al. 2006), the development of flora on the island is highly related to seabird occupation. Observations during our field trips also showed that a great number of trees and shrubs flourished on all these guano-rich islands. The thriving vegetation, in turn, provides an ideal shelter and habitat for seabirds. At the same time, seabirds act as a very effective biological pump, delivering significant quantities of marine-derived nutrients (e.g. N, P, etc.) to their immediate surroundings, and the nutrient-rich guano can greatly improve soil quality and thus enhance bioavailability and plant development. As shown in Figs. 5.1 and 5.2, the source materials of the four profiles gradually changed from a two-component mixture (coral sand and guano) to a three-component system (coral sand, guano and humus) at the respective critical depth in each sediment profile, suggesting the increasing pedogenesis process towards the top sediment layers. Furthermore, the gradual change of organic matter source from guano to humus suggested that large-scale plant development lagged behind the seabirds' occupation of these islands. The four islands followed quite similar ontogenetic patterns, indicating that development of island ecosystems were probably controlled and affected by same driving factor(s). We believe that seabird activities played a vital role in soil development, by making the island ecosystem more favorable to green plants. Seabird-inhabited tropical coral cays received considerable inputs of nutrients N and P from bird guano, which then enhanced nutrient availability, and further contributed to the increase of plant productivity and the development of insular ecosystems (Anderson and Polis 1999; Ellis 2005; Wait et al. 2005; Sigurdsson and Magnusson 2009; Schmidt et al. 2010). The seabird activities appear central to the development of the ecosystems on the Xisha Islands (Polis and Hurd 1996; Polis et al. 1997; Sánchez-Piñero and Polis 2000).

According to the reconstructed seabird population records on Dongdao Island of the Xisha archipelago (Liu et al. 2006), the environment on the tropical coral islands at the early stage of their development was unfavorable to seabirds. Due to the low habitat availability, the island can support only a fragile ecosystem. On this occasion, the island could not support a large number of seabirds, and thus seabird population increased very slowly. As a result, the accumulation of nutrient elements, e.g. N and P, was quite slow (Fig. 5.1). The island ecosystem cannot support much vegetation until the islands had accumulated sufficient bird droppings (and equivalent nutrients). With the continuous accumulation of guano-derived nutrients, metabolic activities and pedogenesis could have been significantly enhanced, ultimately allowing sufficient improvement of sediment and soil structure for plant growth. Such a development pattern for plant community is typical for natural coral islands, where the primary succession process and eventual formation of a climax community (e.g. trees and shrubs) are slow (Eldon and Bradley 2004; Walker and Moral 2003). In addition, as shown in Fig. 5.1, the decrease in P concentrations in the top sediment layer for each of the sediment core contrasted with the increase in levels of organic matter in the bulk sediments, suggesting that the earlier input of nutrient-rich guano had a prolonged and positive effect on the growth of island vegetation, even following a markedly reduced guano input, since the seabird-derived nutrients may remain in soils and sediments for quite a long time, sometimes as long as several hundreds of years (Anderson and Polis 1999; Hawke et al. 1999). Therefore, the organic matter in the upper layers should have mainly stemmed from plants, which is in consistent with our above discussion. Moreover, it is observed from Fig. 5.1 that the organic matter concentrations in the top sediment layers show evident peaks, before decreasing to relatively low levels in the surface 3–4 cm of sediments in cores GJ3, JQ and JY2. This unveils a constant decrease of plant remains input. This may also reflect the recent recession of island ecosystems on the coral islets of the Xisha archipelago due to possible enhanced human perturbation, as well as long-term decrease of guano input, as indicated by the P contents in bulk sediments of the four profiles (Fig. 5.1).

## 5.2 Identification of Avian Bio-elements

Generally, individual elements in sediments are normally affected by complex factors and it may be difficult, or sometimes impossible, to detect individual sources. Despite of this, an assemblage of several elements can be a robust source indicator (Liu et al. 2005; Sun et al. 2000; Huang et al. 2009). As discussed above, the P concentration in the sediments is significantly driven by guano input and thus an ideal marker for seabird activities. A set of elements, including copper (Zu), zinc (Zn), cadmium (Cd), arsenic (As), selenium (Se), barium (Ba), phosphorus (P) and TOC were plotted in Fig. 5.4. From this figure, we noted that the concentration-versus-depth profiles of Cu, Zn, Cd, As, Se, Ba and P displayed quite similar patterns in each of the cores GQ, JQ, JY2 and GJ3 (below 16 cm depth), suggesting that these elements likely sourced from a common material. This conclusion was further confirmed by the Pearson correlation analyses discussed below. The sediments above 16 cm depth of the profile GJ3 had relatively lower P levels (<1 %) than the top sediment samples of the other three profiles, thus the source effect of guano might have been weakened in the upper sediments of GJ3. The plant development and input of plant humus to the coral sediments may further result in the remigration of these elements in the top layer of this profile and this would potentially impact on geochemical behavior of these bio-elements. Consequently, the samples above 16 cm in the core GJ3 were not used in correlation analysis.

The Pearson correlation coefficients between Cu, Zn, Cd, As, Se, Ba and P for the four sediment profiles were given in Table 5.1. All the elements were significantly and positively correlated with P, our indicator of seabird influence. Although As and P in the GJ3 profile showed a relatively low coefficient, the correlation was still significant at the 0.05 level. Except for As in GJ3, correlation between other six elements with P showed quite high degree of confidence ($p < 0.01$), suggesting a common source from seabird faeces. Gong et al. (1996, 1997) studied soils on the tropical islands in the South China Sea, and they found that Cd, Cu, Zn and P were significantly enriched in ornithogenic sediments, and these four elements were positively correlated with each other. Several other related studies have also shown that a few metal elements (e.g. Cd, Se etc.) were enriched in marine animals, for instance seabirds, bivalve and fish etc. (Dietz et al. 1996; Grotti et al. 2008). Thus, we suggest that the assemblage of seven bio-elements, i.e. Cu, Zn, Cd, P, As, Se and Ba, represent important geochemical markers (or bio-marker) for seabird activity on the Yongle archipelago of the Xisha Islands. Furthermore, as shown in Fig. 5.4, with the exception of sediment samples above the critical depths (11, 9,

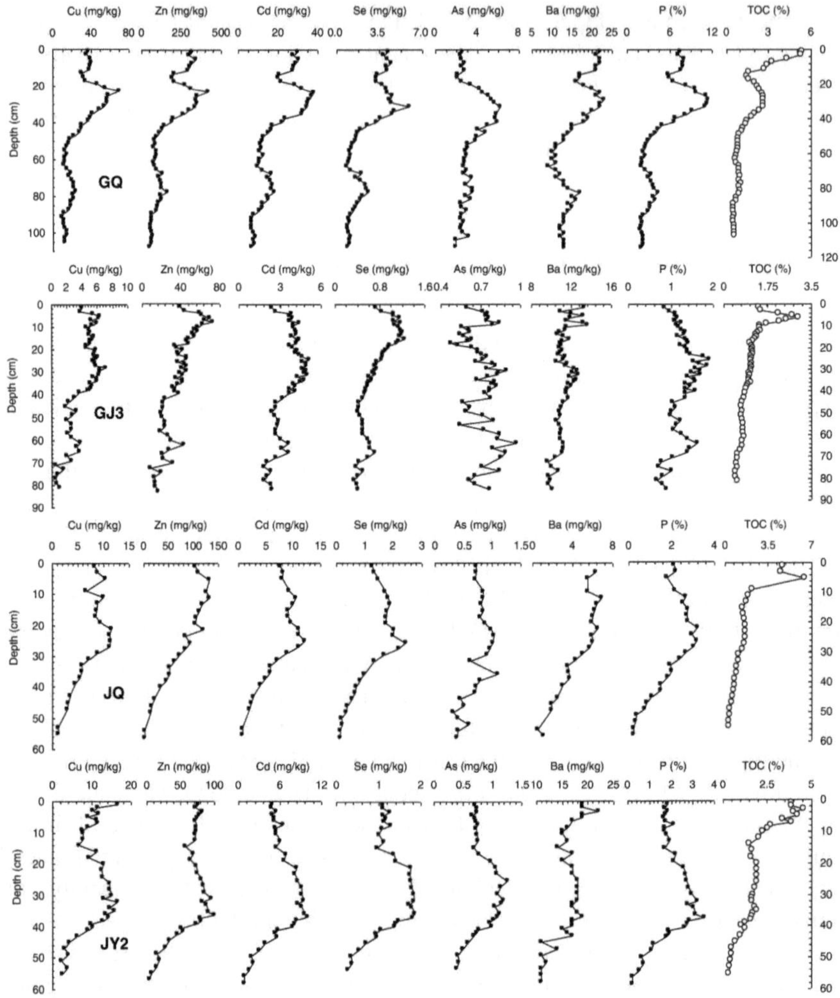

**Fig. 5.4** Down-core profiles of seven elements (Cu, Zn, Cd, Se, As, Ba and P) and TOC in the cores GQ, GJ3, JQ and JY2 (Reprinted from Xu et al. (2011b), Copyright (2011), with permission from Elsevier.)

**Table 5.1** Pearson correlation coefficients between Cu, Cd, Zn, As, S, Ba and P in the four sediment profiles

| | Cu | Cd | Zn | As | Se | Ba |
|---|---|---|---|---|---|---|
| GQ (n = 54) | 0.98** | 0.98** | 0.97** | 0.59* | 0.96** | 0.93** |
| GJ3 (n = 45) | 0.86** | 0.92** | 0.86** | 0.37* | 0.73** | 0.75** |
| JQ (n = 27) | 0.94** | 0.97** | 0.79** | 0.86** | 0.97** | 0.84** |
| JY2 (n = 37) | 0.91** | 0.99** | 0.82** | 0.95** | 0.96** | 0.72** |

** Correlation is significant at 0.01 level; * Correlation is significant at 0.05 level; (Reprinted from Xu et al. (2011b), Copyright (2011), with permission from Elsevier.)

16 and 12 cm for cores GQ, JQ, GJ3 and JY2, respectively), the vertical profiles of Cu, Zn, Cd, P, As, Se and Ba for the four cores mirrored those of TOC, further supporting the predominant contribution of guano to the composition of organic matter in the sediments below the critical depth for each of the four profiles. This also implies that the recent increase in TOC is not caused by an increase in seabird population, but rather more likely an increase in plant development.

From our previous study of the Dongdao Island in the Xuande group of the Xisha Islands, an assemblage of 7 elements (i.e. Cu, Cd, As, Se, Zn, P and S) in the bird-influenced sediments were clustered into one group, and their levels were primarily subject to seabird guano input (Liu et al. 2006). By comparison, it is found that the avian bio-elements, except for Ba, present in GQ, GJ3, JQ and JY2 profiles were the same as those of the Dongdao Island. The absence of Ba as a bio-element on the Dongdao Island may be explained by its post-depositional geochemical changes. The ecological environments of Ganquan, Guangjin, Jinqing and Jinyin Islands are relatively simple, pedogenesis effect on these islands is weaker relative to Dongdao Island, thus the post-depositional alterations in the sediments were insignificant. However, the ornithogenic lake sediments collected from the Cattle Pond in the Dongdao Island were possibly have experienced more complex physical/chemical processes and this might have greatly impacted on the geochemical behavior of Ba (Liu et al. 2006; Xu et al. 2011), leading to its weakened source effect. Sun et al. (2000) investigated geochemical features of ornithogenic sediments from the Antarctica, and has suggested that Ba is also a typical bio-elements in penguin guano fossils. Thus, the group of elements including Cu, Cd, Zn, P, As and Se could act as a "bio-elements" indicator for seabird droppings on all the Xisha Islands, although there are minor differences between the western and eastern Xisha islands. Thus, these elements in the ornithogenic sediments from the tropical coral islands can be served as independent geochemical markers for seabird activity. According to Sun et al. (2000, 2001; Sun and Xie 2001), the changes of bio-element concentrations in the sediments affected by penguin droppings can be used to estimate penguin population size in the past. For the coral sediments influenced by seabird excrement, the levels of bio-elements as discussed above may also be applied to evaluate records of seabird population changes during the late-Holocene. Thus, the concentrations of Cu, Cd, Zn, P, As, Se and Ba have the potential to be used as geochemical markers for tracking seabird paleoecological records in the Xisha archipelago.

The above results are consistent with our theory that seabirds living on the coral islands have transfered marine-derived nutrients to the nutrient-poor coral island ecosystems. Moreover, as with nutrients, seabird droppings are also significant source of metal pollutants such as As, Cd, Cu, Zn from ocean to island environment as a result of biomagnification and bioaccumulation throughout marine food chains. Scientists examined the distribution characteristics of trace elements (Cd, Cu, Zn etc.) and nutrients in the marine environment, and found significant positive correlations between these metal contaminants and nutrients, for example phosphorus (Bruland 1980). Their data support our hypothesis that these trace elements have a common source in ocean. Our finding is in parallel with the results of numerous studies of biovectors in remote islands of the high-latitude Arctic and Antarctica (Zale 1994; Headley 1996; Wagner and Melles 2001; Sun et al. 2000, 2001; Sun and Xie 2001; Michelutti et al. 2009, 2010; Brimble et al. 2009a, b; Blais et al. 2005, 2007). Similarly, Hawke et al. (1999) and Hawke (2003) reported that the seabird occupation and guano accumulation could result in Cd contamination in New Zealand agricultural soils. Our study of ornithogenic sediments from the Xisha Islands adds to the growing body of evidence that biotransfer can be an important pathway of contaminant transport between oceanic ecosystems and terrestrial food webs in tropical islands.

## 5.3 Geochemical Characteristic of Elements Fe, Al, Ti, Mn and K

Many scientists have shown great conerns to the guano fertilizer and vegetation on the Xisha Islands (Exploration Group of Xisha Islands of Institute of Soil Science of Chinese Academy of Sciences 1977; Exploration Group of Xisha Islands of Plant Institute of Guangdong province 1977; Hainan Ocean Administration 1999), and have found that guano input had marked influences on soil development and plant growth on these islands. Except for the bio-elements discussed above, the input of biological remains into the coral islands could also have accelerated the biogeochemical cycling rates of some other elements, exerting significant effects on their distributions, migration and transformation.

Generally, Fe, Al, Ti, Mn, K are lithophile elements and they are normally major elements in sediments which have a terrestrial origin. Down-core profiles of Fe, Al, Ti, Mn, K and TOC are supplied in Fig. 5.5, showing that concentrations of these elements in the sediments are less than 1 % (i.e. equivalent 100 mg/kg). Thus they are identified as trace elements in these coral-derived sediments. Furthermore, as shown in this figure, despite some minor fluctuations in individual metal contents within the different cores, their values remain at a very similar range. The similar geochemical characteristics of these mineral elements and above-mentioned bio-elements among these islands suggest that the eco-environmental features across the whole Xisha Islands are almost identical. Linear regression analysis between Fe, Al, Ti, Mn and K concentrations in the four sediment profiles and TOC (Fig. 5.6) shows strong correlations between these elements and organic matter, suggesting that the mineral elements in the coral islands and organic matter share a common sink. The differences between these mineral elements and above discussed bio-elements are the degree of affinity with TOC. Fe, Al, Ti, Mn, K and TOC display very similar patterns throughout all the four sediment cores, but the positive correlations between bio-elements (Cu, Zn, Cd, P, As, Se and Ba) and TOC are less pronounced above a critical point: 11, 9, 16 and 12 cm for cores GQ, JQ, GJ3 and JY2, respectively (Figs. 5.4, 5.5). Since the mineral component of the sediment is $CaCO_3$, the cation exchange capacity of the sediment is solely dependent on concentrations of organic matter. In the four sediment profiles, the close linkings between TOC and Fe, Al, Ti, Mn, K imply that these mineral elements were possibly introduced into the ornithogenic sediments

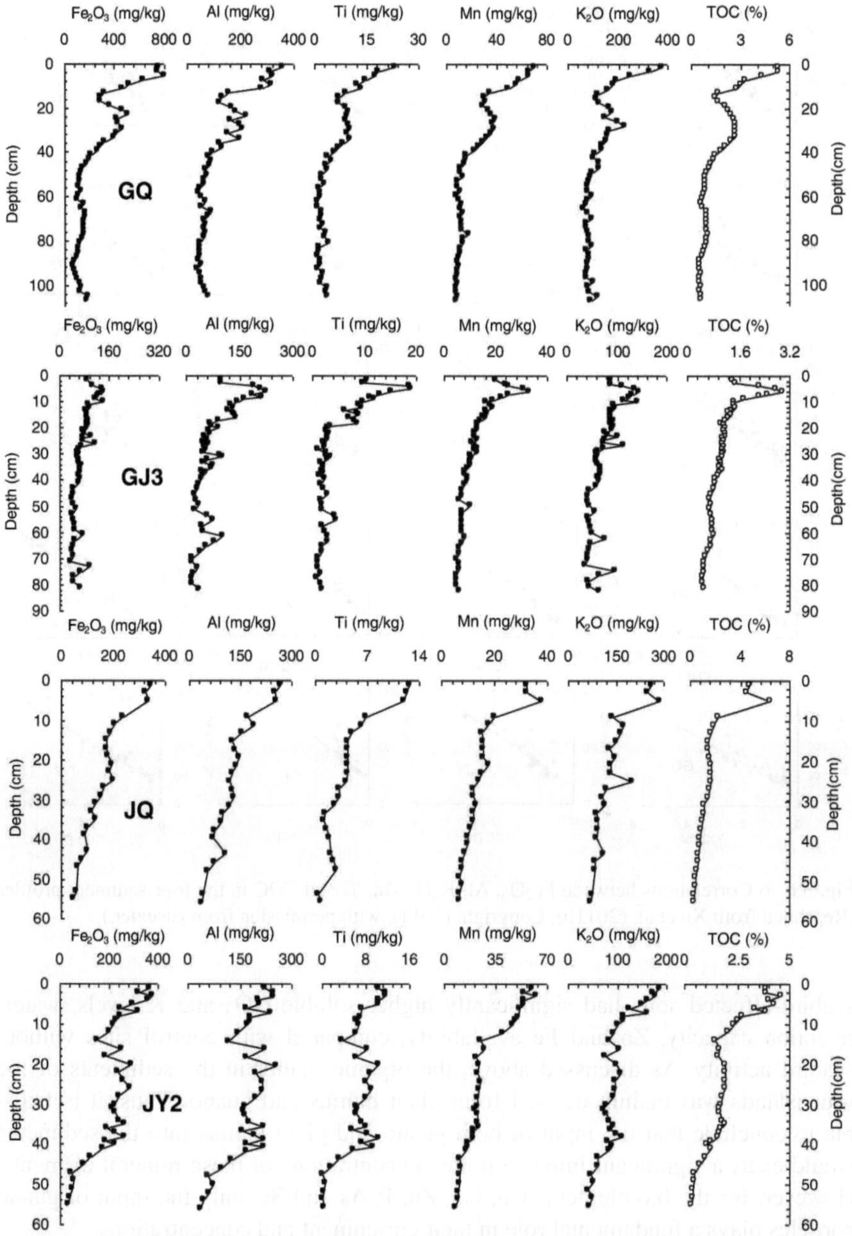


**Fig. 5.5** Concentration-versus-depth profiles of $Fe_2O_3$, Al, Ti, Mn, $K_2O$ and TOC in the sediment cores GQ, GJ3, JQ and JY2 (Reprinted from Xu et al. (2011b), Copyright (2011), with permission from Elsevier.)

through ion exchange with negatively-charged sites on soil organic matter, such that increasing organics abundance results in relatively high contents of these elements in the bulk sediments (García et al. 2002). It was reported that heavily

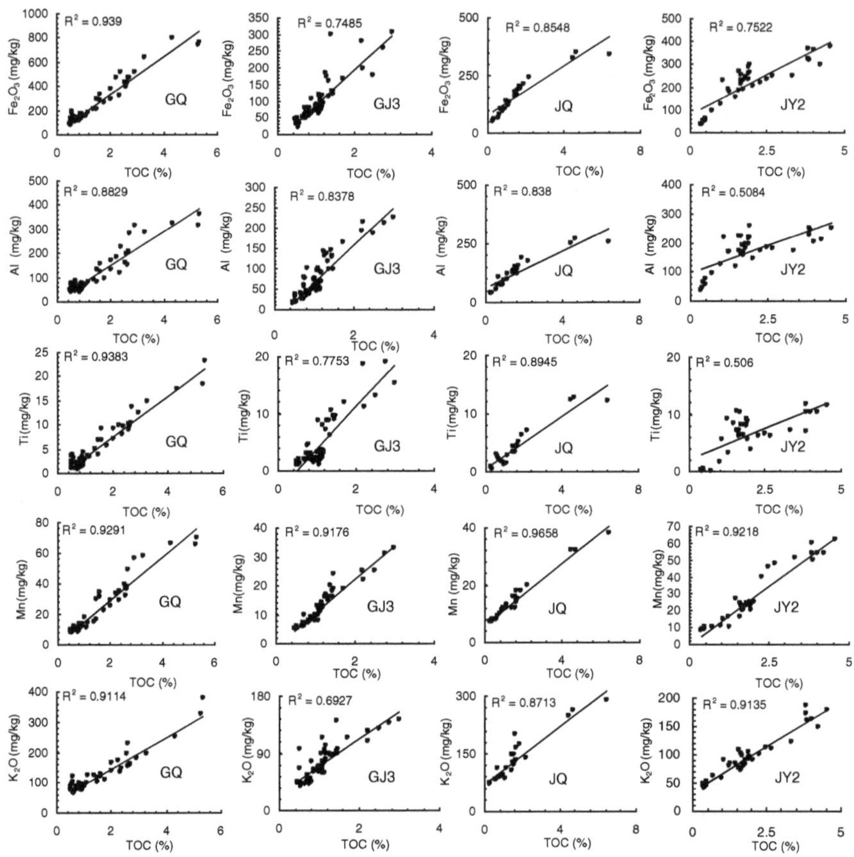


**Fig. 5.6** 6 Correlations between $Fe_2O_3$, Al, $K_2O$, Mn, Ti and TOC in the four sediment profiles (Reprinted from Xu et al. (2011b), Copyright (2011), with permission from Elsevier.)

seabird-affected soils had significantly higher soluble $NO_3$ and K levels, water-retention capacity, Zn and Fe availability, compared with control sites without seabird activity. As discussed above, the organic matter in the sediments of the four islands was mainly derived from plant humus and guano. Thus, it is feasible to conclude that the input of both guano and plant humus into the sediments would exert a significant impact on the concentrations of these mineral elements. However, for the bio-elements Cu, Cd, Zn, P, As and Se, only the input of guano particles plays a fundamental role in their enrichment and concentrations.

We believe there has been a positive and favorable feedback between seabird activities and plant growth on the coral islands. Guano serves as a major nutrient source for plant development and enhances sediment and soil structures. This, in turn, will increase environmental capacity and habitat availability; and further accommodate more seabirds. Such positive interactions elevate metabolic processes, weathering and pedogenesis in the sediments. Indeed, according to our

**Table 5.2** Concentrations of Mn, $Fe_2O_3$, Ti, Al and $K_2O$ in the environmental mediums on the Xisha Islands

| Samples | Mn (mg kg$^{-1}$) | $Fe_2O_3$ (mg kg$^{-1}$) | Ti (mg kg$^{-1}$) | Al (mg kg$^{-1}$) | $K_2O$ (mg kg$^{-1}$) |
|---|---|---|---|---|---|
| Humus (n = 5) | 234.4 | 2448.8 | 85.8 | 1759.1 | 369.0 |
| Leaves (n = 7) | 36.1 | 390.5 | 6.7 | 137.3 | 7,229.8 |
| Coral sand (n = 16) | 18.4 | 67.4 | –[a] | 28.5 | 108.1 |
| Guano (n = 42) | 38.4 | 348.7 | 1.6 | 37.0 | 355.4 |

[a]Too low to be detected (Reprinted from Xu et al. (2011b), Copyright (2011), with permission from Elsevier.)

field lithological observations, the upper layers of the four sediment profiles contained more fine-grained particles than deep layers, and showed a dark-black color. This probably reflects a relatively complex sediment structure above the critical depths of the cores. Organic matter is the most active factor in the process of soil-forming and has a profound capacity of element retaining. We also determined the concentrations of Fe, Al, Ti, Mn and K in individual mediums including humus, plant leaves, guano particles and coral sands. The data are given in Table 5.2, from which we noted that the concentrations of these elements in organic matter are much higher than those in coral sands. Both Fe and Al ions in sediments and soils were readily exchanged with negatively-charged sites on organic matter (Dai 1997; Sparks 2003); moreover, Mn (another exchangeable species) was also found in sediments (Zhang et al. 1998; Akcay et al. 2003). Thus Fe, Al and Mn could be easily adsorbed by organic matter in the soils and sediments. Ti and K probably experienced a similar process, leading to the significant correlation between these elements and organic matter. Furthermore, the correlations of Fe and Al with TOC, and the multiple correlations of some elements with P may also suggest a likely role of Fe/Al-humic-P complexes in binding P. For example, Gerke (2010) reported in their study that Humic-metal-P complexes were highly related to retention of P in soils. An earlier study also provided evidence that a similar process occurred in seabird-impacted soils in New Zealand (Hawke and Powell 1995).

Seabirds have the potential to enhance concentrations of some radionuclides in their immediate habitat, and radionuclide levels were significantly influenced by organic matter input (Dowdall et al. 2003, 2005). On the Xisha islands, although guano is not a source of radionuclides (Refer to Chap. 4 of this book for details), organic matter derived from plant and guano aids in retaining a few radioactive isotopes (Xu et al. 2010). In all, seabird activities, guano accumulation and vegetation development have exerted a profound impact on the texture and composition of sediments and soils in the coral islands. The coral sediments significantly influenced by guano are thus robust natural archives for studying the past eco-environmental changes during the late-Holocene on the islands. The results of the present study construct a geochemical basis for the further reconstruction of seabird population size and plant development history, which is essential for establishing the development mode of the island ecosystems in low-latitude tropical oceans.

## References

Akcay H, Oguz A, Karapire C (2003) Study of heavy metal pollution and speciation in Buyak Menderes and Gediz river sediments. Water Res 37:813–822

Anderson WB, Polis GA (1999) Nutrient fluxes from water to land: seabirds affect plant nutrient status on Gulf of California islands. Oecologia 118:324–332

Baker JC, Jell JS, Hacker LF, Baublys KA (1998) Origin of recent insular phosphate rock on a coral Cay-Ralne Island, Northern Great Barrier Reef, Australia. J Sediment Res 68:1001–1008

Blais JM, Kimpe LE, McMahon D, Keatley BE, Mallory ML, Douglas MSV, Smol JP (2005) Arctic seabirds transport marine-derived contaminants. Science 309:445

Blais JM, Macdonald RW, Mackay D, Webster E, Harvey C, Smol JP (2007) Biologically mediated transport of contaminants in aquatic systems. Environ Sci Technol 41:1075–1084

Brimble SK, Blais JM, Kimpe LE, Mallory ML, Keatley BE, Douglas MSV, Smol JP (2009a) Bioenrichment of trace elements in a series of ponds near a northern fulmar (*Fulmarus glacialis*) colony at Cape Vera, Devon Island. Can J Fish Aquat Sci 66:949–958

Brimble SK, Foster KL, Mallory ML, Macdonald RW, Smol JP, Blais JM (2009b) High Arctic ponds receiving biotransported nutrients from a nearby seabird colony are also subject to potentially toxic loadings of arsenic, cadmium, and zinc. Environ Toxicol Chem 28(11):2426–2433

Bruland KW (1980) Oceanographic distrituions of cadmium, zinc, nickel, and copper in the North Pacific. Earth Planet Sci Lett 47(2):176–198

Dai SG (1997) Environmental chemistry. Higher Education Press, Beijing, pp 210–214

Dietz R, Riget F, Johansen P (1996) Lead, cadmium, mercury and selenium in Greenland marine animals. Sci Total Environ 186:67–93

Dowdall M, Gerland S, Lind B (2003) Gamma-emitting natural and anthropogenic radionuclides in the terrestrial environment of Kongsfjord, Svalbard. Sci Total Environ 305:229–240

Dowdall M, Gwynn JP, Gabrielsen GW, Lind B (2005) Assessment of elevated radionuclide levels in soils associated with an avian colony in a high Arctic environment. Soil Sediment Contam 14:1–11

Eldon DE, Bradley FS (2004) Environmental science: a study of interrelationships. Tsinghua University Press, Beijing, pp 108–120

Ellis JC (2005) Marine birds on land: a review of plant biomass, species richness, and community composition in seabird colonies. Plant Ecol 181:227–241

Exploration Group of Xisha Islands of Institute of Soil Science of Chinese Academy of Sciences (1977) Soil and guano phosphorus mine in Xi-Sha Islands. Science Press, Beijing (in Chinese with English abstract)

Exploration Group of Xisha Islands of Plant Institute of Guangdong province (1977) Plants and vegetation on Xisha Islands of China. Science Press, Beijing (in Chinese with English abstract)

García LV, Marañón T, Ojeda F, Clemente L, Redondo R (2002) Seagull influence on soil properties, chenopod shrub distribution, and leaf nutrient status in semi-arid Mediterranean islands. Oikos 98:75–86

Gerke J (2010) Humic (Organic Matter)-Al(Fe)-Phosphate Complexes: An Understimated Phosphate Form in Soils and Source of Plant-Available Phosphate. Soil Sci 175(9):417–425

Gong ZT, Liu LW, Zhou RR (1996) Formation of the soils in the South China Sea islands and their ages. Quat Sci 1:88–95 (in Chinese with English abstract)

Gong ZT, Huang B, Zhou RR (1997) Pedogeochemistry and bioavailability of soils in South China Sea Islands. Acta Pedol Sin 34(1):10–27 (in Chinese with English abstract)

Grotti M, Soggia F, Lagomarsino C, Riva SD, Goessler W, Francesconi KA (2008) Natural variability and distribution of trace elements in marine organisms from Antarctic coastal environments. Antarct Sci 20(1):39–51

Hainan Ocean Administration (1999) The professional proceedings of the integrated investigation research on sea islands resource of Hainan province. China Ocean Press, Beijing (in Chinese with English abstract)

Hawke DJ (2003) Cadmium distribution and inventories at a pre-European seabird breeding site on agricultural land, Banks Peninsula. New Zealand. Aust. J. Soil Res 41:19–26

Hawke DJ, Powell HKJ (1995) Soil solution chemistry at a Westland petrel breeding colony, New Zealand-paleoecological implications. Aust J Soil Res 33:912–924

Hawke DJ, Holdaway RN, Causer JE, Ogden S (1999) Soil indicators of pre-european seabird breeding in New Zealand at sites identified by predator deposits. Aust J Soil Res 37:103–113

Headley AD (1996) Heavy metal concentrations in peat profiles from the high Arctic. Sci Total Environ 177:105–111

Huang T, Sun LG, Wang YH, Liu XD, Zhu RB (2009) Penguin population dynamics for the past 8500 years at Gardener Island, Vestfold Hills. Antarct Sci 21(6):571–578

Liu XD, Sun LG, Zhao SP, Yin XB, Xie ZQ, Luo HH (2005) Eco-environmental information recorded in the lake sediments of the Dongdao Island, South China Sea. Quat Sci 25(5):574–584 (in Chinese with English abstract)

Liu XD, Zhao SP, Sun LG, Luo HH, Yin XB, Xie ZQ, Wang YH, Liu KX, Wu XH, Ding XF, Fu DP (2006) Geochemical evidence for the variation of historical seabird population on the Dongdao Island of South China Sea. J Paleolimnol 36:259–279

Liu XD, Sun LG, Cheng ZQ, Zhao SP, Liu KK, Wu XH, Xie ZQ, Yin XB, Luo HH, Ding XF, Fu DB, Wang YH (2008) Paleoenvironmental implications of the guano phosphatic cementation on Dongdao Island in the South China Sea. Mar Geol 247:1–16

Michelutti N, Keatley BE, Brimble S, Blais JM, Liu H, Douglas MSV, Mallory ML, Macdonald RW, Smol JP (2009) Seabird-driven shifts in Arctic pond ecosystems. Proc R Soc B 276:591–596

Michelutti N, Blais JM, Mallory ML, Brash J, Thienpont J, Kimpe LE, Douglas MSV, Smol JP (2010) Trophic position influences the efficacy of seabirds as metal biovectors. Proc Natl Acad Sci 107(23):10543–10548

Polis GA, Hurd SD (1996) Linking marine and terrestrial food webs: allochthonous input from the ocean supports high secondary productivity on small islands and coastal land communities. Am Nat 147:396–423

Polis GA, Anderson WB, Holt RD (1997) Towards an integration of landscape and food web ecology: the dynamics of spatially subsidized food webs. Annu Rev Ecol Syst 29:289–316

Sánchez-Piñero F, Polis GA (2000) Bottom-Up dynamics of allochthonous input: direct and indirect effects of seabirds on islands. Ecology 81(11):3117–3132

Schmidt S, Mackintosh K, Gillett R, Pudmenzky A, Allen DE, Rennenberg H, Mueller JF (2010) Atmospheric concentrations of ammonia and nitrogen dioxide at a tropical coral cay with high seabird density. J Environ Monit 12:460–465

Sigurdsson BD, Magnusson B (2009) Ecosystem respiration, vegetation development and soil nitrogen in relation to breeding density of seagulls on a pristine volcanic island, Surtsey, Iceland. Biogeosciences Discuss 6:8393–8409

Sparks DL (2003) Environmental Soil Chemistry. Academic Press, London, pp 101–109

Sun LG, Xie ZQ (2001) Relic: penguin population programs. Sci Prog 83(1):31–44

Sun LG, Xie ZQ, Zhao JL (2000) A 3,000-year record of penguin populations. Nature 407:858

Sun LG, Xie ZQ, Zhao JL (2001) The sediments of lake on the Ardley Island, Antarctica: Identification of penguin-dropping soil. Chin J Polar Sci 12(1):1–8

Wagner B, Melles M (2001) A Holocene seabird record from Raffles Sø sediments, East Greenland, in response to climatic and oceanic changes. Boreas 30:228–239

Wait DA, Aubrey DP, Anderson WB (2005) Seabird guano influences on desert islands: soil chemistry and herbaceous species richness and productivity. J Arid Environ 60:681–695

Walker LR, Moral RD (2003) Primary succession and ecosystem rehabilitation. Cambridge University Press, London, pp 50–63

Wang GZ (2001) Sedimentology of the coral reef in South China Sea. Ocean Press, Beijing, pp. 1-310 (in Chinese)

Xu LQ, Liu XD, Sun LG, Yan H, Liu Y, Luo YH, Huang J, Wang YH (2010) Distribution of radionuclides in the guano sediments of Xisha Islands, South China Sea and its implication. J Environ Radioact 101:362–368

Xu LQ, Liu XD, Sun LG, Yan H, Liu Y, Luo YH, Huang J (2011a) A 2 200-year record of seabird population on Ganquan Island, South China Sea. Acta Geol Sin-Engl Ed 85(4):957–967
Xu LQ, Liu XD, Sun LG, Yan H, Luo YH, Huang J (2011b) Geochemical evidence for the development of coral island ecosystem in the Xisha Archipelago of South China Sea from four ornitogenic sediment profiles. Chem Geol 286:135–145
Zale R (1994) Changes in size of the Hope Bay Adele Penguin rookery as inferred from lake Boeckella sediments. Ecography 17:297–304
Zhang CS, Zhang S, Wang LJ, Wang LZ (1998) Comparison of geochemistry of metals in sediments from some typical rock areas. Acta Sci Circumst 18(2):172–176

# Chapter 6
# Reconstruction of Seabird Population Record on the Xisha Islands

Seabird population size in a specific ecosystem is a typical marker for environment quality. If the relative ratio of guano in the bulk sediments can be obtained, the records of seabird population will be reconstructed. In addition, reconstruction of seabird density is essential for a better understanding of the relationships between lives and the environment in the Xisha Islands. We have establish the chronology for the four profiles by combination of radiocarbon and $^{210}$Pb dating (Refer to Chap. 4 for details). In this chapter, we will attempt to reconstruct the record of seabird population on Ganquan Island over the past 2,000 years by geochemical and spectral analysis of the ornithogenic sediments, and examine the possible mechanism of seabird population changes in response to climate change and human activity.

## 6.1 Reconstruction of Seabird Population on the Ganquan Island Over the Past 2,200 Years

According to $^{14}$C dating, the core GQ dates back to approximately 2,200 years, much older than the other four cores. This renders it the best archive in the five cores to study seabird palaeoecology. As discussed in Chap. 5 and our previous study (Liu et al. 2006), seven elements (Cu, Zn, Cd, P, As, Se and S) in sediments of Dongdao Island have been identified as avian bio-elements, and they could be used to track past seabird population history on Dongdao Island. On this basis, The seabird population and vegetation development history on Dongdao Island between the time of 1,350–350 year BP, a period without significant human activities, had been reconstructed. Furthermore, Sr/Ca and Mg/Ca ratios of the three source materials of the sediments differ significantly, based on which the relative contributions of seabird droppings, coral sands and plant remains to the lacustrine



L. Xu, *Impact of Climate Change and Human Activity on the Eco-environment*,
Springer Theses, DOI 10.1007/978-3-662-45003-1_6

sediments were calculated using a three end-member mixing model. The development process of island ecosystem over the past 1,800 years was then established (Zhao et al. 2007). For the GQ profile, the down core distributions of Cu, Cd, Ba, Se, As, Zn and P displayed very similar patterns (Fig. 6.1). It is observed that their two peak zones at respective layers of 20–40 and 70–90 cm were in accordance with those of TOC contents (Fig. 6.1). Regression analysis also reveals that these elements are positively and significantly correlated with P (Fig. 6.2), suggesting that they originate from a common source. Based on elemental geochemistry analysis of the ornithogenic sediments from Dongdao Island and Ganquan Island, several elements Cu, Zn, Cd, P, As, Se and Ba are highly linked with guano input and thus their assemblage is an ideal bio-marker for relative population size of seabird.

We performed Principle Component Analysis (PCA) on the concentration profiles of Cu, Zn, Cd, Ba, As, Se and P. Principal Component 1, i.e. PC1, explains

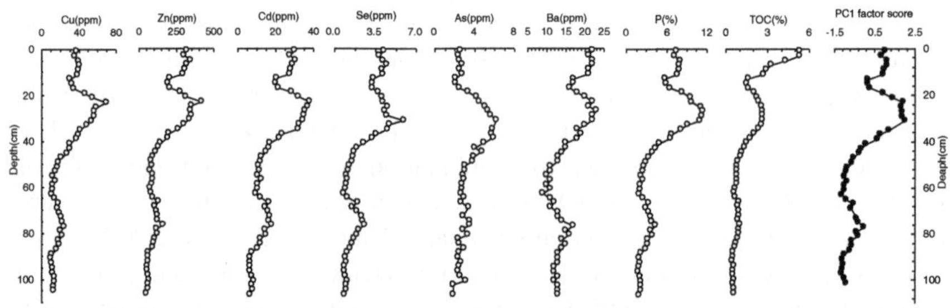


**Fig. 6.1** Down-core variation profiles of seven inorganic elements, TOC and primary component PC1 score value for the seven chemical elements in the GQ sediments core

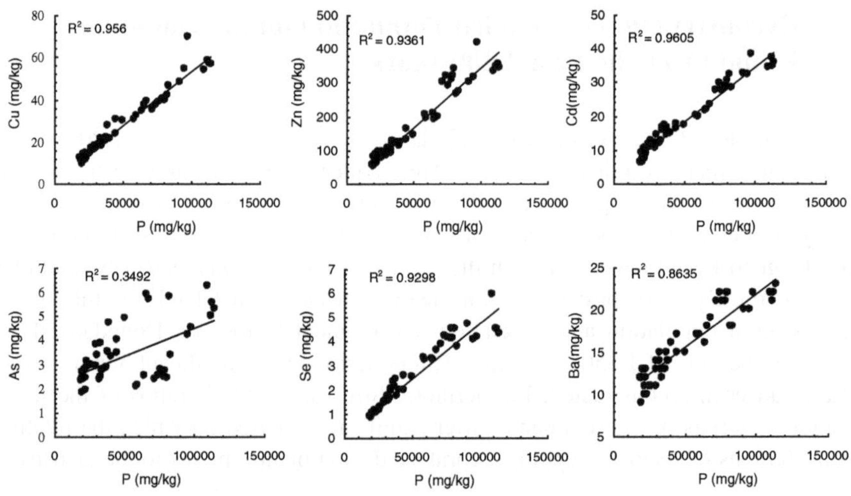


**Fig. 6.2** Correlations between Cu, Zn, Cd, Ba, As, Se and P in the sediments of GQ. Correlation is significant at the 0.01 level

86.59 % of the total variance, and is thus the driving factor on the concentrations of these seven elements. As discussed above and in Chap. 5, Cu, Zn, Cd, Ba, As, Se and P were principally derived from seabird faeces, PC1 is thus most likely linked with the guano input. This is further confirmed by the fact that profile of PC1 mirrors that of P (Fig. 6.1). Therefore, PC1 could be used as a proxy for guano input and seabird population on the Ganquan Island.

The reconstructed record of seabird population changes on the Ganquan Island as indicated by PC1 is given in Fig. 6.3. From this figure, it is clear that seabirds began to inhabit the Ganquan Island as a habitat more than two millennia ago. In general, the seabird population history showed two evident peaks between the respective periods of 2,100–1 850 year BP and 900–300 year BP. Seabird population reached a small peak at approximately 2,100 year BP, after that it decreased gently. Since 1,850 year BP, the seabird population was featured by long-term and slow increase; it then reached the main peak at ~600 year BP; and afterwards, the population showed a rapid decrease. A marked trough of seabird population size appeared at the time between 300 and 200 year BP; and in the recent century, the population seemed to decrease slightly.

The characteristics of the reconstructed record of seabird population on the Ganquan Island are summarized as follows: (1) The increase in population is gradual and slow, but its decrease can be rather fast. This strictly conformed to the natural law for development of seabird population: the changes of environmental factors, for instance abrupt climate change, disease and food shortage, could readily lead to abrupt decrease in bird population, but its enlargement and recover is generally quite slow (Newton 1998). (2) The size of seabird population during the peak time of 900– 300 year BP was evidently larger than that during the

**Fig. 6.3** Comparisons between seabird population records of Ganquan Island and Dongdao Island. *A* record of seabird population on the Dongdao Island using avian bio-elements (Liu et al. 2006); *B* PC1 values extracted from bio-elements, indicating seabird population change of the Ganquan Island, the present study; *C* Seabird population record on the Dongdao Island deduced from Sr/Ca and Mg/Ca (Zhao et al. 2007)

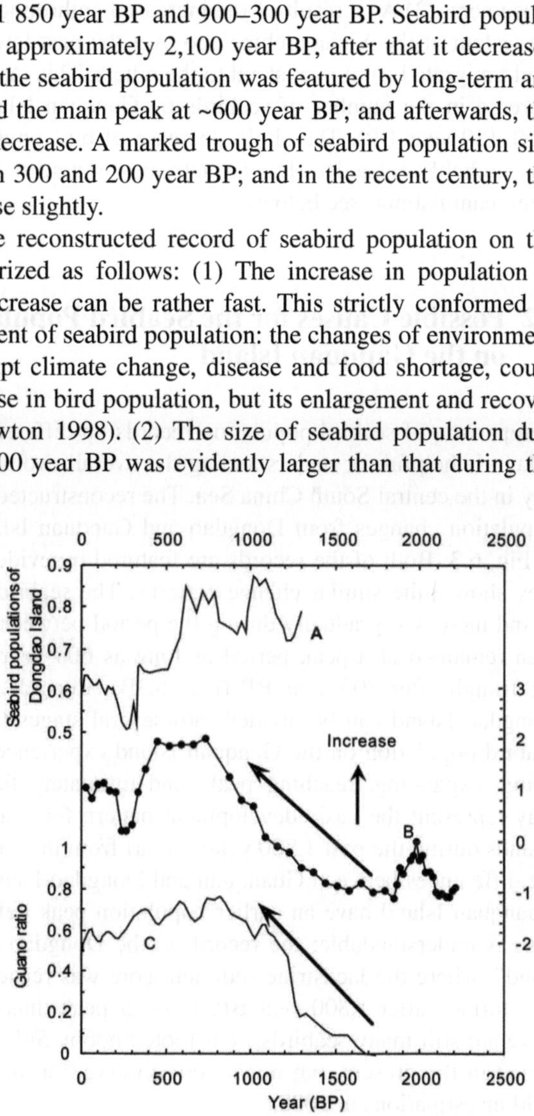

period of 2,100–1,850 year BP; and the peak duration was longer, suggesting that environmental capacity of the island at different times is likely the driving factor controlling seabird populations on a small island. This is consistent with our previous research (Zhao et al. 2007). When the island was initially formed, there was little seabird-dependent plant on this small island, and such a flora system was very fragile with low resilience, seabirds could easily suffered from severe disasters, and a large number of seabirds cannot be supported. During the period of 900–300 year BP, it is very likely that the vegetation developed and remains at a relatively stable level. This increased environmental capacity, i.e. habitat availability, for seabirds, and the island could thus accommodate more top seabird consumers. (3) As a result of the increased anthropogenic influences on the island ecosystem of the Xisha archipelago over the past few hundred years, the seabirds might migrated from one island to the others. This may ultimately lead to the rapid increase in the number of seabirds on Ganquan Island within a very short time (200–150 year BP). The slight decrease of bird population over the past century was probably related to human activities around the Yongle group, for example significant fishing (see below).

## 6.2 Possible Causes for the Seabird Population Alterations on the Ganquan Island

Comparison of seabird population records on different islands of the Xisha archipelago is helpful for understanding the overall evolution patterns of seabird ecology in the central South China Sea. The reconstructed records of historical seabird population changes from Dongdao and Ganquan Islands are plotted respectively in Fig. 6.3. Both of the records are featured by evident fluctuations. Nonetheless, they show quite similar change patterns. The seabird population on the Ganquan Island increased gradually during the period between 1,700 and 900 year BP, and then remained at a peak period as long as 600 years before a sharp decrease to the trough after 300 year BP (Fig. 6.3B). Record of seabird population on the Dongdao Island can be divided into several stages (Fig. 6.3A, C). Likewise, the seabird population on the Ganquan Island experienced four similar phases: occupying, expanding, reaching peak, and ultimately falling down. Such a process may represent the basic development pattern for seabird population in the Xisha Islands during the past 1,800 years. Apart from the similarities, there are two notable differences between Guanquan and Dongdao Islands. First, the seabirds on the Guanquan Island have an earlier population peak before the point 1,800 year BP. This is understandable: the record on the Dongdao Island was based on "Cattle Pond", where the lacustrine sediment core was retrieved and analyzed. The pond was formed after 1,800 year BP, an older peak thus cannot be recorded. Second, there are still many seabirds, Red-footed booby *Sula Sula*, living on the Dongdao Island in the present, but none were observed on the Ganquan Island during our field investigations in 2008.

The coherent similarities of seabird population records from different islets suggest that the seabird ecology in the Xisha archipelago was probably subject to same environmental and ecological factor(s). To investigate the possible reasons for seabird population changes, we plot several major climatic factors available for the Xisha archipelago, including Sea Surface Temperature (SST) in the west Pacific, three temperature records (from Greenland Arctic, China and northern hemisphere, respectively), grain size compositions of the China Yellow Sea sediments indicative of the East Asian winter monsoon strength, and redness of lake sediments from the Qinghai lake in the Tibet Plateau (Fig. 6.4). To decode the potential links between the seabird population changes on the Ganquan Island and earth surface temperature, we compared the reconstructed seabird population record with widely-accepted temperature records from the Greenland ice core (Fig. 6.4A) and the Qinghai lake (larger values of redness corresponds to greater

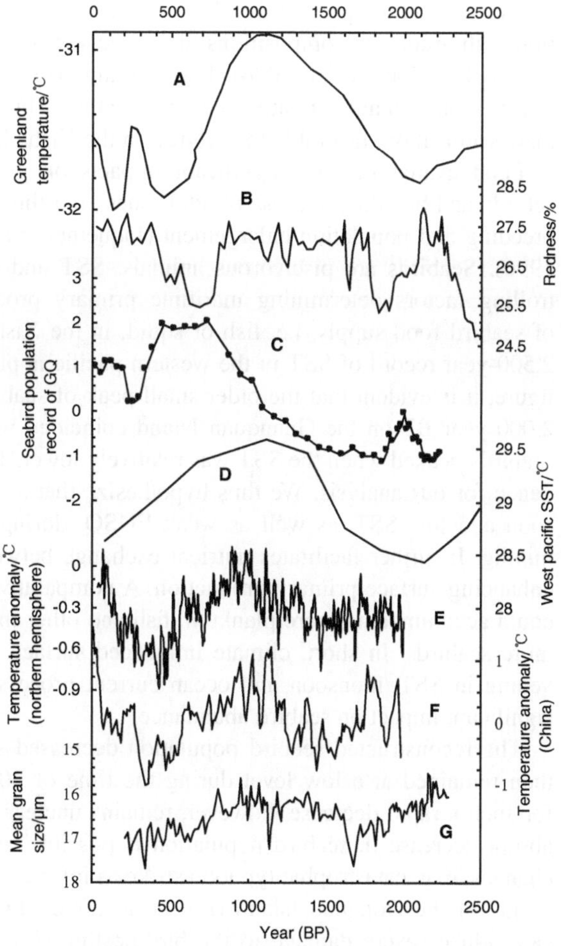


**Fig. 6.4** Comparisons between seabird population changes on the Ganquan Island and climatic records. Cited literature: *A* Temperature record of Greenland (Dahl-Jensen et al. 1998); *B* Redness record of Qinghai lake (Ji et al. 2005); *C* PC1 factor score extracted from bio-elements, indicating seabird population changes in Ganquan Island; *D* SST record of western Pacific Ocean (Stott et al. 2004); *E* Temperature record of Northern Hemisphere (Moberg et al. 2005); *F* Temperature record of China (Yang et al. 2002); *G* Grain size record of China Yellow Sea (Xiang et al. 2006)

atmospheric temperature Fig. 6.4B, Ji et al. 2005). We noted a marked association. The period of population peak between 900 and 300 year BP (Fig. 6.4C) is almost in consistent with the time of the climate transition from the Medieval Warm Period (WMP), or Medieval Climate Anomaly (MCA), to the Little Ice Age (LIA). We also compared the seabird population record with the proxy-based surface temperature histories of northern hemisphere (Fig. 6.4E, Moberg et al. 2005) and China (Fig. 6.4F, Yang et al. 2002). The recent seabird population peak did not appear during the Medieval Warm Period (1,100–800 year BP). In contrast, the seabird population increased when the climate was relatively cool, and the number of seabirds peaked during the Little Ice Age (LIA, 600–100 year BP, Cronin et al. 2003). The open sea around the Xisha Islands was characterized by a relatively dry and cool climate during the LIA. For example, Wei et al. (2004) studied high-resolution Sr/Ca and Mg/Ca ratios of coral Porites lutea from the Xisha Islands and found that the monthly-averaged summer SSTs in seawater around the Xisha Islands approximately 540 years ago (the Little Ice Age) were more than 1℃ lower than the present day. We also compared the PC1-based seabird population with grain size compositions of marine sediments from the China Yellow Sea (Fig. 6.4G, Xiang et al. 2006). Larger grain size reflected a stronger East Asian winter monsoon and a relative low temperature. Thus, we conclude that a cool climate seems more favorable to seabirds on the Xisha Islands.

Food availability has significant impacts on growth rate of young juvenile sebirds and breeding success of adult ones, and thus plays a critical role in birds' breeding and population enlargement (Jaquemet et al. 2004; Megyesi and Griffin 1996). Seabirds are piscivorous animals. SST and nutrient supply are the controlling factors determining maritime primary productivity and the abundance of seabird food supply, i.e. fish or squid, in the Xisha archipelago. A continuous 2,500-year record of SST in the western pacific is plotted in Fig. 6.4D. From this figure, it is evident that the older small peak of seabird population approximately 2,000 year BP on the Guanquan Island coincided with low SST. The number of seabirds peaked when the SST was relatively lower. This further provide more evidence for our analysis. We thus hypothesize that a strong East Asia winter monsoon and low SST, as well as weak ENSO, during cool times induced oceanic mixing. It further facilitates nutrient exchange between surface and deep waters, enhancing surface primary production. A comparatively high surface productivity could accommodate more plankton, fish and other marine organism, and supports more seabirds. In short, climate influenced surface primary production by intervening in SST, monsoon and ocean current processes, which ultimately exerted significant impact on seabird abundance.

The reconstructed seabird population decreased sharply at ~300 year BP and then remained at a low level during the time of 300–200 year BP. The reasons for such a rapid decrease, however, remains unclear so far. We anticipate that the abrupt decrease in seabird population is possibly related to rapid environmental changes, e,.g. catastrophic typhoon strike as inferred from the sediments in "Cattle Pond" of the Dongdao Island (Liu et al. 2006). Firstly, such catastrophic event can induce severe damage to the bird-nesting shrubs and woodlands. Secondly, the gale and heavy rainfall brought about by the intense typhoon activity could directly destroy breeding birds, young chicks, and eggs. All these processes are fatal to predator seabirds.

The ornithogenic sediments retrieved from Ganquan Island, Xisha archipelago, contains important information about seabird palaeoecology in the past. In short, a set of seven elements, including Cu, Zn, Cd, Ba, As, Se and P, were identified as avian bio-elements. The assemblage of those elements was an ideal geochemical marker for the influence of guano input on the ornithogenic sediments, and the concentrations of elements were highly dependent on seabird guano input. The occupation history of seabird on the Ganquan Island during the last 2,200 years was further reconstructed by principal component analysis (PCA) on these seven bio-elements. Despite of the remarkable fluctuations, the reconstructed seabird population on the Ganquan island displayed two notable peaks at the periods of 2,100–1,850 and 900–300 year BP, respectively. By comparing of the reconstructed seabird population records on the Dongdao Island and the Ganquan Island, it is found that the seabird population records from two individual islands showed quite similar change patterns. The history seabirds' occupation ofthe Xisha Islands can be divided into four phases: occupying, slowly expanding, keeping peak and rapidly falling. This represents a basic development mode of seabird ecology. The relationship between the seabird population changes over the last two millennia on the Ganquan Island and climatic variables seems complex. It is found that a relatively dry and cold climate is more favorable to tropical seabirds. Further research is required to reconstruct marine primary productivity, so as to better understand detailed responses of seabirds to climate changes.

## 6.3 Reflectance Spectroscopy: A New Approach to Reconstructing Seabird Population

### *6.3.1 Introduction to Reflectance Spectroscopy*

There are a great number of geographically isolated islands, e.g. the Xisha archipelago, in the South China Sea. These pristine islands have not received significant disturbance and thus provide useful information for island development, evolution of species and environmental changes, as well as ecological changes of island ecosystems in response to climate changes. So far, the long-term interactions between seabird occupation and plant development on these small islands has not been well understood. The gradual development of such isolated island ecosystems remains unclear. Changes in source material compositions of the bulk sediments are essential for identifying the development of seabird-inhabited islands and potential environmental stress on these islands. Thus, an analysis of the compositions and source materials of the widely distributed ornithogenic sediments on such coral islands is indispensable. However, traditional chemical assay is normally time consuming, sample destructive and quite expensive. Developing an effective and rapid approach to reconstructing levels of source materials will be helpful for obtaining a rapid overview of the sedimentary process and identifying potential ecological and environmental implications.

Due to its convenience, accuracy and rapidity, reflectance spectrum within the visible-near-infrared (Vis-NIRS) region, i.e. 380–2,500 nm has been widely used to predict chemical structures of sediments, soils and biological remains (Malley and Williams 1997; Rosén et al. 2000; Wu et al. 2005; Roumet et al. 2006; Hay et al. 2010; Showers et al. 2006). As well, near-infrared reflectance spectroscopy (NIRS) can be utilized to quantify the ecological properties of large sample numbers, and thus this simple technique can offer enormous analytical power for ecologists (Kleinebecker et al. 2009; Foley et al. 1998; Stolter et al. 2006). In our earlier study, the reflectance spectra of ornithogenic sediments from the Antarctic have been investigated, and we successfully estimated bio-element levels, indicative of relative penguin population size, using a two-component model (Liu et al. 2010, 2011). While our previous studies was established on a basis of two-component system, whether this approach is applicable to more complicated three-component mixtures in the low-latitude Xisha Islands remains a problem. One of our aims in the present study is to examine the feasibility of using spectroscopy to predict source material levels on a three-component basis. We examine the potential application of reflectance spectroscopy to reconstruct the relative contributions of source materials, i.e. coral sand, guano and plant humus, to the coral-based ornithogenic sediments from the Xisha Islands, South China Sea. Furthermore, we also attempt to investigate possible environmental and ecological implications of changes in source material levels reconstructed by reflectance spectroscopy in latter part of this chapter.

### *6.3.2 Analytical Methods*

As stated in preceding chapters, endmembers of the bulk sediments, including coral sand, guano and plant humus, were collected in situ during field investigation. Prior to reflectance spectral analysis, the manually collected plant remains were treated with excess 5 % hydrochloric acid in a 50 ml centrifuge tube under an ultrasonic environment. The acid was changed at regular time-intervals, until any possible carbonate was thoroughly removed. The plant humus was then repeatedly washed by de-ionized water, so as to remove excess hydrochloric acid and it was then air dried in our laboratory. The bulk sediments and the three source materials were processed as follows: subsamples source materials were homogenized with a mortar and pestle after being dried to a constant weight under a temperature of 105 °C. These samples were then passed through a stainless steel sieve (200 mesh). The fine-grained sample was placed into a standard measuring cell. The diffuse reflectance spectrums of the samples were then obtained on a recording spectrophotometer (Shimadzu DUV-3700 UV-Vis-NIR) by scanning at a range from 380 to 2,500 nm, at intervals of 1 nm. Thus, 2,121 spectral data were acquired for an individual sample. To estimate relative concentrations of source materials, the spectrums of three endmembers (i.e. guano, plant humus and coral sand) were also analyzed. For quality assurance (QA) and quality control (QC), an external reference standard, zero absorbance polyethylene, was read alternately within every batch of the samples. The obtained reflectance data (r) were then recalculated as absorbance (− log r) data using software UV. Prove.

### 6.3.3 Seabird Population Reconstruction

Large quatities of well-preserved seabird droppings (guano particles) were observed throughout all the four sediment profiles. Consequently, the influence of guano would have a certain influence on the spectral property of the bulk sediments. As discussed in earlier chapters, those islands were manufactured by reef-building corals, and coral sands are thus the most principal constituent in our ornithogenic sediments. In addition to coral sand and guano particles, both laboratory analyses and field observations have shown that plant remain is also an important component of the bulk sediments (Xu et al. 2010). Thus, we assume that guano, plant humus and coral sand are three dominant parental materials for the sediments on the Xisha Islands. The physical and chemical properties of these three source materials are very different, which enables a potential use of reflectance spectrum to reconstruct their relative ratios in the bulk sediments. The determined absorbance spectra data of the three components between 380 and 2,500 nm are given in Fig. 6.5. In our earlier study, spectral properties of Antarctic ornithogenic sediments have been investigated, based on which we estimated relative penguin population size, using a two-component mixing model. The result from reflectance spectrums was consistent with that obtained from penguin bio-element analyses (Liu et al. 2010, 2011). This further validated the reliability of using reflectance technique to estimate penguin population size. As to the ornithogenic sediments from the Xisha archipelago, significant differences among the absorbance spectra (1/r) (380–2,500 nm) for the three source materials were observed, and spectrums for each of the three endmembers seems quite stable (Fig. 6.5). This is helpful for identifying the signal from each endmember. Due to the difference in spectral pattern, we suggest that the relative contributions of these three components to the bulk sediments can be reconstructed on a three-endmember basis. The spectral pattern of subsample is a combination of the characteristic spectrums from three endmembers. Using the software Matlab (version 7.1), the ratio of each component is calculated from the equation below:

$$A = X \cdot A_{guano} + Y \cdot A_{plant} + Z \cdot A_{coral}$$

where $A_{guano}$, $A_{plant}$ and $A_{coral}$ are specific absorption spectrums (380–2,500 nm) of guano, plant humus and coral sand. X, Y and Z in the equation are respectively the hypothesized ratio of each component. They conform to the condition $X + Y + Z = 1$. At any wavelength from 380 to 2,500 nm, $A_{guano}$, $A_{plant}$ and $A_{coral}$

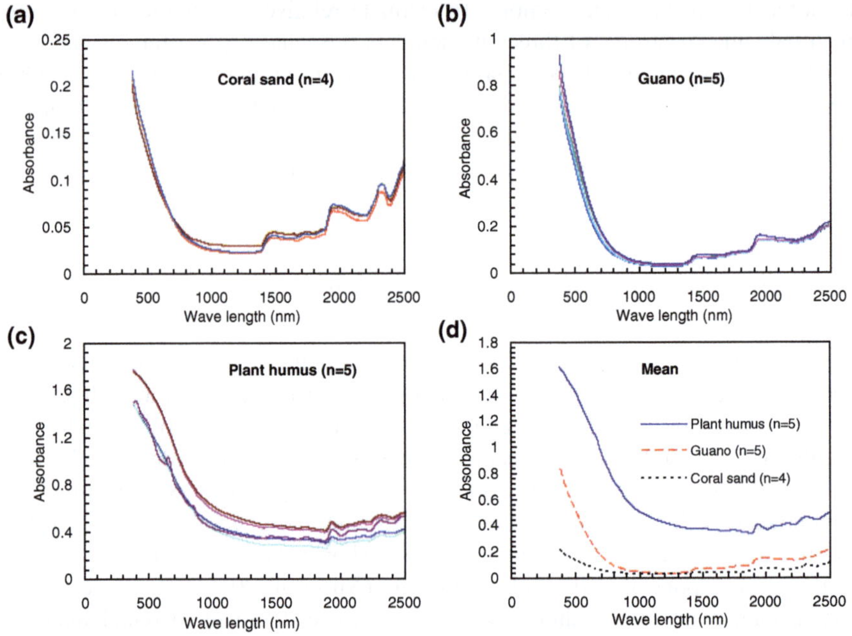


**Fig. 6.5** Characteristic spectra of three endmembers. **a–c** are spectra of different environmental mediums, and **d** is the mean of each (reprinted from Xu et al. (2012), Copyright (2012), with permission from Elsevier)

in the equation were constant. We assigned different values to X, Y and Z at increments of 1 % for each variable, and obtained a total of $10^4$ spectral curves, plotting the plant, guano and coral sand in the range from 0 to 100 %. For each of the spectrums, we hypothesize that X, Y and Z are the estimated ratios of guano, plant humus and coral sands in our bird-affected sediments when the calculated A best fits sample's spectral pattern $A_{sample}$ (tested by least square principle). Using the equation shown above, relative ratios of plant, guano and coral sand in each sample of the sediment cores GQ, JQ, GJ3 and JY2 were reconstructed. The results were supplied in Fig. 6.6.

Coral debris is the pedogenic parent material for the anthropogenic sediments on the Xisha archipelago, and thus $CaCO_3$ is the dominant constituent. To examine reliability of our approach to reconstructing guano/plant/coral ratios, we performed Pearson correlation analyses on the coral percentages reconstructed by physical reflectance spectroscopy and the level of $CaCO_3$ contents determined by chemical analysis for the whole profiles. The carbonate concentrations in each profile and the results of correlation analysis are given in Fig. 6.7. The coral sand ratios are positively and significantly correlated with $CaCO_3$ contents ($R > 0.85$, $p < 0.01$) for all the four sediment profiles.

Concentration of phosphorus in ornithogenic sediments is a robust marker for seabird influence (Sun et al. 2000; Xu et al. 2010; Huang et al. 2009). In general,

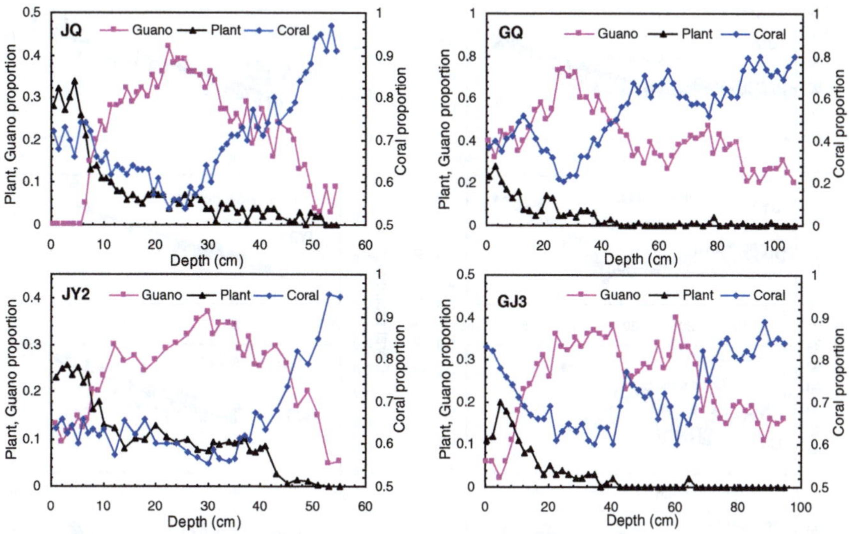


**Fig. 6.6** Down-core distributions of guano, coral sand and plant proportions on Jinqing, Jinyin, Ganquan and Guangjin islands (reprinted from Xu et al. (2012), Copyright (2012), with permission from Elsevier)

the distributions of phosphorus as a function of depth is consistent with guano levels that derived from spectrum analysis (Fig. 6.8), with minor differences for the top sediments in each profile as shown by the shaded area in Fig. 6.8. The spectrum-based guano proportions in the sediments below the critical depths of the four profiles (11, 9, 16 and 7 cm for cores GQ, JQ, GJ3 and JY2, respectively) and phosphorus concentrations yield a significant positive correlation (Fig. 6.8). In contrast, the correlation between guano proportions and phosphorus levels in the top layers of each profile seems less pronounced. Such a discrepancy between guano ratio and phosphorus in the top sediments of each profile can be explained by gradual vegetation development and further migration of nutrient element phosphorus. According to both field study and published data, seabird population density on the Xisha Islands has significantly declined in the past century (Cao et al. 2007). The guano ratio in recent sediments reconstructed from reflectance spectrum has also decreased abruptly (Fig. 6.8). In that nutrients are essential for plant development, nutrient-containing guano in top layer of the sediments could be absorbed by plants. The consumption of guano by vegetation is probably the reason for the decrease of guano levels in parallel to seabird population regression. Since the level of nutrient phosphorus in guano (~15 %) is much greater than that in plant (<0.5 %), a very small part of bird faeces can therefore support a large number of vegetation. Thus, we conclude that consumption of guano by island flora may not significantly decrease the guano level in the bulk sediments, and guano ratios reconstructed by reflectance spectrum can serve as an indirect marker for relative size of seabird population.

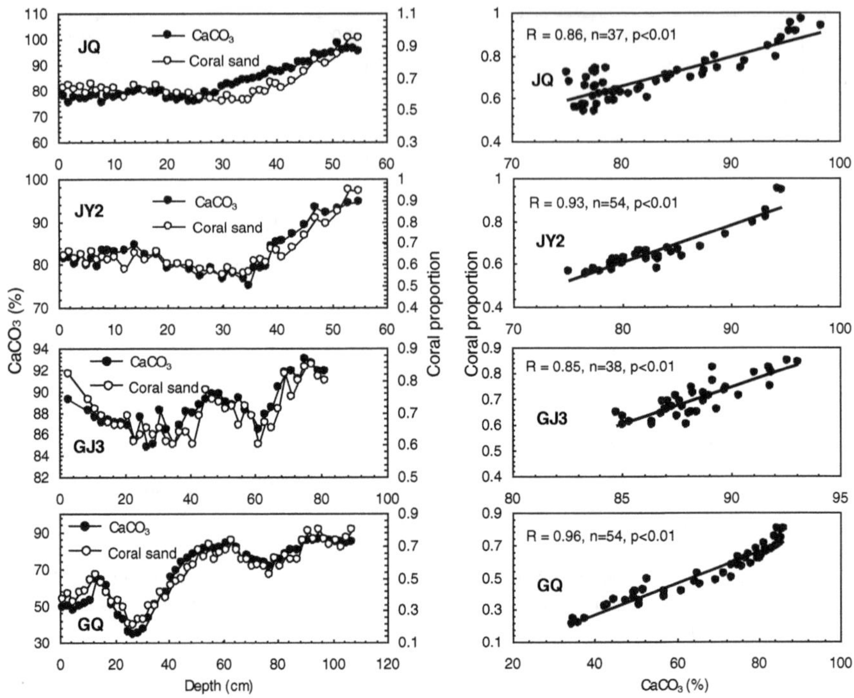


**Fig. 6.7** Correlations between coral proportions from reflectance spectroscopy and $CaCO_3$ levels in the profiles JY2, JQ, GJ3 and GQ (reprinted from Xu et al. (2012), Copyright (2012), with permission from Elsevier.)

Guano ratios show rapid decrease in upper layers of each profile, the concentration of phosphorus in the bulk sediments, however, seems constant in top layers of the sediments (Fig. 6.8, Shaded area). This discrepancy is also possibly caused by vegetation development on these islands. Although seabird abundance in the last century has decreased substantially, nutrient-rich guano acts as a ideal nutrient-pool that continues to support vegetation development on the islands. The development of plants may have lead to the remigration of some avian bio-elements, for example phosphorus, and these bio-elements would be further enriched in plant-derived organic matter. Indeed, it is found that contents of phosphorus in plant leaves and humus on the islands are comparatively high (Table 6.1). In terms of seabird population reconstruction, reflectance spectroscope performs better than chemical analysis. Thus, we conclude that reflectance spectroscope is a rapid, non-destructive and effective approach to reconstructing seabird occupation history and plant development process on these coral islands.

According to $^{210}Pb$ test, the abrupt decrease in seabird population began at approximately 1850 AD. Such a rapid decrease is probably caused by human activities. For example, human and introduced species decrease habitat availability for seabirds (Cao 2005). Overfishing led to a sharp decrease in fish resources

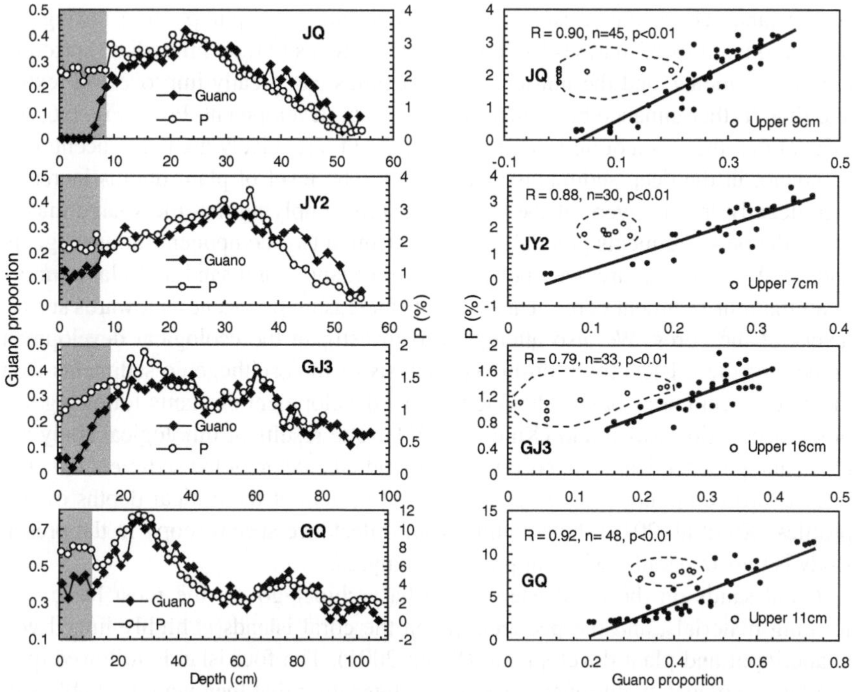


**Fig. 6.8** Correlations between guano proportions from reflectance spectroscopy and phosphorus levels in the four profiles (shaded area: upper 11, 16, 9 and 7 cm for cores GQ, GJ3, JQ and JY2, respectively) (reprinted from Xu et al. (2012), Copyright (2012), with permission from Elsevier)

**Table 6.1** Phosphorus level in plants and humus

| Sample | Plant A | Plant D | Humus 1 | Humus 2 | Humus 3 | Humus 4 | Humus 5 |
|---|---|---|---|---|---|---|---|
| P (mg $kg^{-1}$) | 1,095 | 3,987 | 25,652 | 18,239 | 70,769 | 2,901 | 3,046 |

Reprinted from Xu et al. (2012), Copyright (2012), with permission from Elsevier

and further seabird population decline. Furthermore, ground water and soils on the Xisha Islands have been severely contaminated (Xie et al. 2005). This may also be partly responsible for decrease in seabird population.

### *6.3.4 Significance for the Development of Seabird Island Ecosystem*

As demonstrated in earlier chapters, these ornithogenic sediment cores experienced little disturbance and have been well-preserved. Thus the changes in source material compositions can tell us critical information for the development history of coral island ecosystems. Seabirds are effective biological pumps, transferring significant quantities of marine-derived nutrient elements to their immediate surrounding environments, and the nutrient-rich droppings can greatly improve soil texture and thus further enhance bioavailability and plant development. From Fig. 6.6, it is evident that there is a quite low level of plant, but a relatively greater concentration of guano, in the deep sediments of the cores. The level of plant humus increased significantly in top layers of the cores (Fig. 6.6), implying that the source materials of the bulk sediments gradually changed from a two-component mixture (guano and coral sand) to a three-component system (guano, coral sand and plant humus) in all the four sediment cores. This suggests increasing pedogenesis towards surface layers of the cores. We also attempted to investigate the ecological development process of these islets by multi-proxy analysis of the ornithogenic sediments from these sediment cores, based on which a slow development of plants following seabirds' occupation was found (Xu et al. 2011). The results of lithological study and elemental geochemical analysis revealed a gradual change of the sediment structure from a two-component to a three-component system at the critical depths of four profiles (Xu et al. 2011). Our finding from reflectance spectroscopy in the present study is thus in parallel with our previous analyses.

Coral sands on the coral islets of Xisha archipelago are the most basic soil-forming materials, and the pedogenesis on the coral islands is highly linked with guano input and plant development (Wang 2001). The four islands followed quite similar ecological evolutionary processes, suggesting that they were probably subject to similar driving factor(s). We believe that seabird communities and their activities played a central role in sediment and soil developments on these islands, by making them more favorable for vegetation development. Seabird islands can receive significant inputs of phosphorus (P) and nitrogen (N) from bird droppings, which then enhanced nutrient level and bioavailability, leading to remarkable increase in primary productivity and further development of isolated ecosystems (Anderson and Polis 1999; Ellis 2005; Sigurdsson and Magnusson 2009; Schmidt et al. 2010). Briefly, seabird activities are central to the ecosystem development of the Xisha archipelago.

The hypothesized development of the coral island ecosystems in the South China Sea is given in Fig. 6.9. According to Charles Darwin's theory, ocean coral islands are volcanic in origin. At the initial phase, reef building corals began to grow around extinct marine volcanic island and ultimately "manufactured" reefs and coral islands. Such islands were extremely lacked in nutrients for plant development. We believe that, after the islands were formed, seabirds began to settle on these insular islands, and this was perhaps the most important event during the development of island ecosystems. Seabirds created the nutrient pathways, transferring considerable quantities of nutrients from ocean to islands (Allaway and Ashford 1984; Anderson and Polis 1999). When sufficient nutrients, such as phosphorus and nitrogen, were accumulated, the islands became favorable to vegetation.

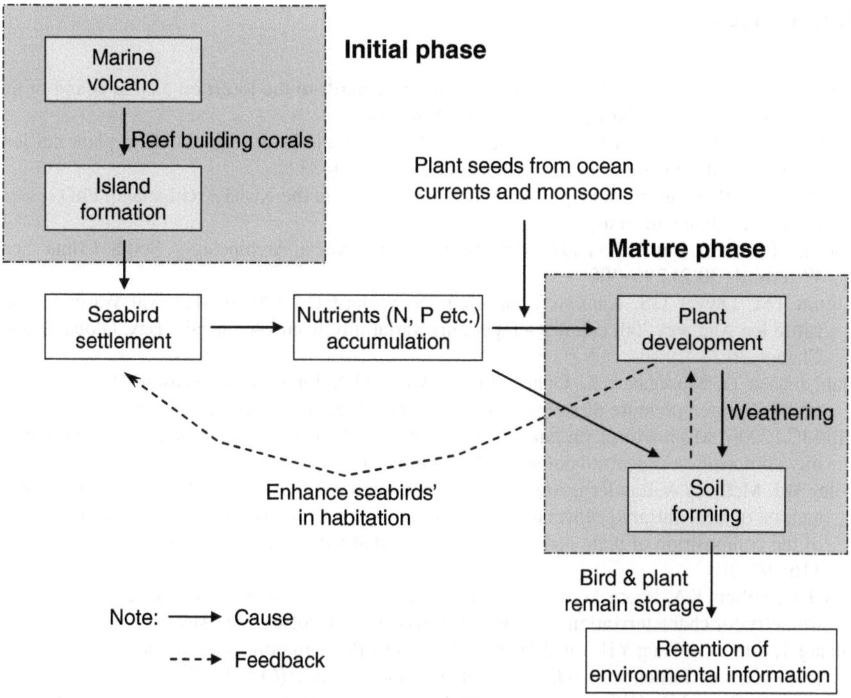


**Fig. 6.9** Schematic diagram of the development of coral island ecosystem in the Xisha Archipelago

The original plant seeds probably originated from Hainan Island: the similarity between plants in the Xisha archipelago and that on the large Hainan Island is as high as 91 % (Exploration Group of Xisha Islands of Plant Institute of Guangdong province 1977). Plant seeds may further have been transported to the islands through ocean currents or tsunami. The mature phase of island ecosystems begins with vegetation development. Seabirds brought the required nutrients for plants, and the vegetation, in turn, provided habitats for seabirds. Vegetation and bird droppings enhanced weathering and soil development. This further aids storage of bird and plant remains, which, otherwise, would probably be flushed into the ocean by surface runoff.

Environmental information was ultimately retained in such natural archives. For example, several naturally occurring isotopes brought back to the earth surface by precipitation, and anthropogenic radionuclides have been detected in such ornithogenic sediments (See Chap. 4) and we will reconstruct contamination history of heavy metals from well-preserved ornithogenic sediments (See Chap. 8 for details).

## References


Allaway WG, Ashford AE (1984) Nutrient input by seabirds to the forest on a coral island of the Great Barrier Reef. Mar Ecol Prog Ser 19:297–298

Anderson WB, Polis GA (1999) Nutrient fluxes from water to land: seabirds affect plant nutrient status on Gulf of California islands. Oecologia 118:324–332

Cao L (2005) Population ecology of the red-footed booby on the Xisha Archipelago. PhD dissertation. Lanzhou University

Cao L, Pan YL, Liu NF (2007) Waterbirds of the Xisha Archipelago, South China Sea. Waterbirds 30(2):296–300

Cronin TM, Dwyer GS, Kamiya T, Schwede S, Willard DA (2003) Medieval Warm Period, Little Ice Age and 20th century temperature variability from Chesapeake Bay. Global Planet Change 36:17–29

Dahl-Jensen D, Mosegaard K, Gundestrup N, Clow GD, Johnsen SJ, Hansen AW, Balling N (1998) Past temperature directly from the Greenland ice sheet. Science 282:268–271

Ellis JC (2005) Marine birds on land: a review of plant biomass, species richness, and community composition in seabird colonies. Plant Ecol 181:227–241

Foley WJ, Mcllwee A, Lawler I, Aragones L, Woolnough AP, Berding N (1998) Ecological applications of near infrared reflectance spectroscopy—a tool for rapid, cost-effective prediction of the composition of plant and animal tissues and aspects of animal performance. Oecologia 116:293–305

Hay KB, Millers KA, Poore AGB, Lovelock CE (2010) The use of near infrared reflectance spectrometry for characterization of brown algal tissue. J Phycol 46:937–946

Huang T, Sun LG, Wang YH, Liu XD, Zhu RB (2009) Penguin population dynamics for the past 8500 years at Gardener Island, Vestfold Hills. Antarct Sci 21(6):571–578

Jaquemet S, Corre ML, Weimerskirch H (2004) Seabird community structure in a coastal tropical environment: importance of natural factors and fish aggregating devices (FADs). Mar Ecol-Prog Ser 268:281–292

Ji JF, Shen J, William B, Chen J, Liu LW, Liu XQ (2005) Asian monsoon oscillations in the northeastern Qinghai-Tibet Plateau since the late glacial as interpreted from visible reflectance of Qing hai Lake sediments. Earth Planet Sci Lett 233:61–70

Kleinebecker T, Schmidt SR, Fritz C, Smolders AJP, Hölzel N (2009) Prediction of $\delta^{13}C$ and $\delta^{15}N$ in plant tissues with near-infrared reflectance spectroscopy. New Phytol 184:732–739

Liu XD, Zhao SP, Sun LG, Luo HH, Yin XB, Xie ZQ, Wang YH, Liu KX, Wu XH, Ding XF, Fu DP (2006) Geochemical evidence for the variation of historical seabird population on Dongdao Island of the South China Sea. J Paleolimnol 36:259–279

Liu XD, Sun J, Sun LG, Liu WQ, Wang YH (2010) Feasibility of using reflectance spectroscopy for the analysis of bio-element concentrations in Antarctic ornithogenic sediments. Chin J Polar Sci 21(1):31–45

Liu XD, Sun J, Sun LG, Liu WQ, Wang YH (2011) Reflectance spectroscopy: a new approach for reconstructing penguin population size from Antarctic ornithogenic sediments. J Paleoliminol 45:213–222

Malley DF, Williams PC (1997) Use of near-infrared reflectance spectroscopy in prediction of heavy metals in freshwater sediment by their association with organic matter. Environ Sci Technol 31:3461–3467

Megyesi JL, Griffin CR (1996) Breeding biology of the Brown Noddy on Tern Island, Hawaii. Wilson Bull 108(2):317–334

Moberg A, Sonechldn DM, Holmgren K, Datsenko NM, Karlen W (2005) Highly variable Northern Hemisphere temperatures reconstructed from low- and high-resolution proxy data. Nature 433:613–617

Newton I (1998) Population limitation in birds. Academic Press, London, pp 10–15

Rosén P, Dåbakk E, Renberg I, Nilsson M, Hall R (2000) Near-infrared spectrometry (NIRS): a new tool for inferring past climatic changes from lake sediments. Holocene 10(2):161–166

Roumet C, Picon-Cochard C, Dawson LA, Joffre R, Mayes R, Blanchard A, Brewer MJ (2006) Quantifying species composition in root mixtures using two methods: near-infrared reflectance spectroscopy and plant wax markers. New Phytol 170:631–638

Schmidt S, Mackintosh K, Gillett R, Pudmenzky A, Allen DE, Rennenberg H, Mueller JF (2010) Atmospheric concentrations of ammonia and nitrogen dioxide at a tropical coral cay with high seabird density. J Environ Monito 12:460–465

Showers SE, Tolleson DR, Stuth JW, Kroll JC, Koerth BH (2006) Predicting diet quality of white-tailed deer via NIRS fecal profiling. Rangeland Ecol Manage 59:300–307

Sigurdsson BD, Magnusson B (2009) Ecosystem respiration, vegetation development and soil nitrogen in relation to breeding density of seagulls on a pristine volcanic island, Surtsey, Iceland. Biogeosciences Discuss 6:8393–8409

Stolter C, Julkunen-Tiitto R, Ganzhorn J (2006) Application of near infrared reflectance spectroscopy (NIRS) to assess some properties of a sub-arctic ecosystem. Basic Appl Ecol 7:167–187

Stott L, Cannariato K, Thunell R, Haug GH, Koutavas A, Lund S (2004) Decline of surface temperature and salinity in the western tropical Pacific Ocean in the Holocene epoch. Nature 431:56–59

Sun LG, Xie ZQ, Zhao JL (2000) A 3,000-year record of penguin populations. Nature 407:858

Wang GZ (2001) Sedimentology of the coral reef in South China Sea. Ocean Press, Beijing (in Chinese)

Wei GJ, Yu KF, Li XH, Zhao JX, Sun YL, Sun M, Nie BF (2004) Coralline Sr/Ca and Mg/Ca thermometer for the northern South China Sea: calibration and primary application on high resolution SST reconstructing. Quat Sci 24(3):325–331

Wu YZ, Chen J, Ji JF, Tian QJ, Wu XM (2005) Feasibility of reflectance spectroscopy for the assessment of soil mercury contamination. Environ Sci Technol 39:873–878

Xiang R, Yang ZS, Saito Y, Guo ZG, Fan DJ, Li YH, Xiao SB, Shi XF, Chen MH (2006) East Asia winter monsoon changes inferred from environmentally sensitive grain-size component records during the last 2300 years in mud area southwest off Cheju Island, ECS. Sci Chin Ser D Earth Sci 36(7):652–662

Xie ZQ, Sun LG, Zhang PF, Zhao SP, Yin XB, Liu XD, Cheng BB (2005) Preliminary geochemical evidence of groundwater contamination in coral islands of Xi-Sha, South China Sea. Appl Geochem 20(10):1848–1856

Xu LQ, Liu XD, Sun LG, Yan H, Luo YH, Huang J, Wang YH (2010) Distribution of radionuclides in the guano sediments of Xisha Islands, South China Sea and its implication. J Environ Radioactiv 101:362–368

Xu LQ, Liu XD, Sun LG, Yan H, Liu Y, Luo YH, Huang J (2011) Geochemical evidence for the development of coral island ecosystem in the Xisha Archipelago of South China Sea from four ornithogenic sediment profiles. Chem Geol 286:135–145

Xu LQ, Liu XD, Sun LG, Liu WQ (2012) Rapid identification of source materials in coral sand ornithogenic sediments by reflectance spectroscopy. Ecol Indicators 23:517–523

Yang B, Braeuning A, Johnson KR, Shi YF (2002) General characteristics of temperature variation in China during the last two millennia. Geophys Res Lett 29(9):38-1–38-4

Zhao SP, Sun LG, Liu XD, Xie ZQ, Luo HH, Liu KX, Wu XH, Ding XF, Fu DP (2007) Using Sr/Ca and Mg/Ca ratios source indicators of ornithogenic lacustrine sediments on coral island. Quat Sci 27(1):149–156 (in Chinese with English abstract)

# Chapter 7
# Isotopic Evidence for Seabird Diet Changes Over the Past 2000 Years on the Xisha Islands

## 7.1 Introduction to Stable Isotope Ecology

Stable isotopes have been frequently used as tracers in physical and biological systems, and this has made them increasingly important to ecological studies. For example, stable isotope compositions in a variety of environmental mediums may differ significantly, rendering a widely use of stable isotope techniques to examine trophic positions, dietary compositions and foraging behaviors of animals, especially seabirds (Hobson 1999; Post 2002). "Stable isotope ecology" has gradually become a new branch of "ecology" in recent years. Stable isotopes of some light elements, i.e., hydrogen, carbon, nitrogen, oxygen, sulfur etc., are widely used in such kind of studies. Of these light elements, carbon and nitrogen isotopes are the most representative tracers. Normally, stable isotopes show evident gradients in an marine environment. For instance, latitudinal, inshore/offshore and planktonic/benthic gradients have been frequently observed. Typically, stable carbon isotope signatures $^{13}C/^{12}C$ relative to a standard (also known as $\delta^{13}C$) have been widely used to estimate foraging behaviors and habitats of predators (Cherel and Hobson 2007; France 1995). A number of studies also demonstrated that stable nitrogen isotope ($\delta^{15}N$) exhibits a stepwise enrichment at each trophic level. In general, $\delta^{15}N$ in the tissues of consumers is approximately 3–5 ‰ (mean 3.4 ‰) greater than those of their diets, and thus serves as an ideal marker for an organism's trophic level (Wade et al. 1991; Post 2002; Bearhop et al. 2004). Scientists investigated carbon and nitrogen isotope compositions in ancient bone collagen samples, and reconstructed changes in the diet of California condors from the Pleistocene to the recent (Chamberlain et al. 2005). Using stable isotopes, Emslie and Patterson (2007) reported abrupt recent shift in $\delta^{13}C$ and $\delta^{15}N$ values in Adélie penguin eggshells from the Antarctica. The results suggested an abrupt shift to lower trophic positions within the past ~200 years. Huang et al. (2011) analyzed seal hairs spanning the last century by $\delta^{15}N$ and found gradual decrease in krill abundance in the Southern



L. Xu, *Impact of Climate Change and Human Activity on the Eco-environment*,
Springer Theses, DOI 10.1007/978-3-662-45003-1_7

Ocean. Statistics on studies of marine mammalian by stable isotopes suggest increasing use of stable isotopes to investigate animals' foraging behavior over the past decades (Newsome et al. 2010), implying that stable isotope study has drawn increasing attention in recent years and gradually become a hot research topic.

Food scarcity would lead to intra/inter-specific competitions, thus we believe that there exists a potential link between seabird abundance and their foraging strategy. Norris et al. (2007) found that abundance of a bird species marbled murrelet was related closely and positively to its trophic position (indicated by feather $\delta^{15}N$) over a period of more than 40 years. In addition the population size of murrelet was significantly and positively correlated with the estimated proportions of fish in the diet. Using stable isotope analysis, Hilton et al. (2006) found that there was evident linking between penguin ecology and the environment. However, they used discrete samples from museums and provided a relatively short-term record in their study. To examine the long-term interaction between seabird population and foraging behavior, a long-term sequential natural archive is required. There were more than 60 seabird species on the islands of the Xisha archipelago (Exploration Group of Xisha Islands of Institute of Soil Science of Chinese Academy of Sciences 1977), but most of them have disappeared from the islands. Only few scientists have studied the seabird ecology on the Xisha Islands in modern times (Cao et al. 2005, 2007), and none of previous studies have investigated seabird foraging behavior in the past and their palaeodiet. From observation data, seabird population decreased significantly within the past century. The reason for such a decrease, however, remains unclear, and we known little about the potential influences of climatic and environmental changes on the foraging behavior and diets of the seabirds so far. The birds on these islands may completely vanish in the near future, and it becomes urgent to assess the vulnerability and resilience of such coral island ecosystems on a scientific basis. Bird relics (feces, bone, eggshell etc.) were observed in the ornithogenic sediments during our field trips, although seabirds have disappeared on most of the islets. In the present chapter, we will attempt to reconstruct isotope-based palaeodiets of the seabirds on the Xisha Islands, so as to examine the possible impacts of climate variability on predator seabirds and their prey.

## 7.2 Analytical Methods

### *7.2.1 Collagen Extraction*

To examine seabird palaeodiets, we determined stable nitrogen and carbon isotope compositions in our ancient collagen samples. Following Longin (1971) and Brown et al. (1988), collagen from bone samples was isolated as follows. Avian bone subfossils were de-mineralized under a weak acid environment (0.5 M HCl) at a temperature of 4 °C for 10 h with acid changed at regular intervals (normally 1 h) until all the minerals were completely dissolved. The samples were then repeatedly rinsed with de-ionized water and centrifuged. To remove humic acids,

0.125 M sodium hydroxide was added to react for 30 min. The mixture was then rinsed again with de-ionized water to neutrality and gelatinized within a weak acidic solution (0.001 M HCl, pH 2.5) at 70 °C for 48 h. The collagen was dissolved in the supernatant, and was ultimately collected, frozen and lyophilized.

### 7.2.2 Stable Isotope Analysis

Stable nitrogen and carbon isotope compositions in ancient collagen samples were analyzed at the University of Ottawa, Canada. Collagen samples, as well as standards, were flashily combusted and the generated gases were adsorbed by a specific "purge and trap" column and sent to Isotope-ratio mass spectrometry (IRMS) interface first and then to IRMS. Both the interface (Conflo II) and IRMS (Delta XP Plus Advantage) were manufactured by Thermo, Germany. Stable isotope compositions of nitrogen and carbon were give in δ-expression, i.e. the deviation from standards in the unit of parts per mil (‰):

$$\delta X(‰) = \left(\frac{R_{sample}}{R_{std}} - 1\right) \times 1000$$

where X is $^{13}C$ or $^{15}N$, and R is the respective $^{13}C/^{12}C$ or $^{15}N/^{14}N$ ratio. The $R_{std}$ values for nitrogen and carbon were based on the atmospheric $N_2$ (AIR) and Vienna Pee Dee Belemnite (V-PDB), respectively. Calibration standards for $\delta^{15}N$ are USGS-40 (accepted value −4.25 ‰), IAEA-N1 (+0.4 ‰), IAEA-N2 (+20.3 ‰) and USGS41 (+41.57 ‰); and $\delta^{13}C$ values are calibrated against USGS-40 (accepted value −26.24 ‰), IAEA-CH-6 (−10.4 ‰), NBS-22 (−29.91 ‰), and USGS-1 (+37.76 ‰). Internal standards are ($\delta^{13}C$, $\delta^{15}N$): C-54 caffeine (−34.46 ‰, −16.61 ‰), C-51 Nicotinamide (−22.95 ‰, 0.07 ‰), C-52 mixture of ammonium sulphate plus sucrose (−11.94 ‰, 16.58 ‰), blind standard C-55: glutamic acid (−28.53 ‰, −3.98 ‰). Replicate measurements of internal standards showed that the precision (Standard deviation) for nitrogen and carbon isotopic measurement is better than ±0.22 ‰ and ±0.12 ‰, respectively.

## 7.3 A 2000-Year Record of Seabird Diets on Ganquan Island

### 7.3.1 Seabird Population Reconstruction

Using reflectance spectroscopy (For details, refer to Chap. 6), we reconstructed a 2200-year record of seabird population on Ganquan Island (Fig. 7.1c). We also reconstructed the historical seabird abundance by bio-element analysis Chap. 6. Figure 7.1a supplied time-series of the avian bio-elements, and the seabird population record reconstructed from these elements was shown in Fig. 7.1b. There is no

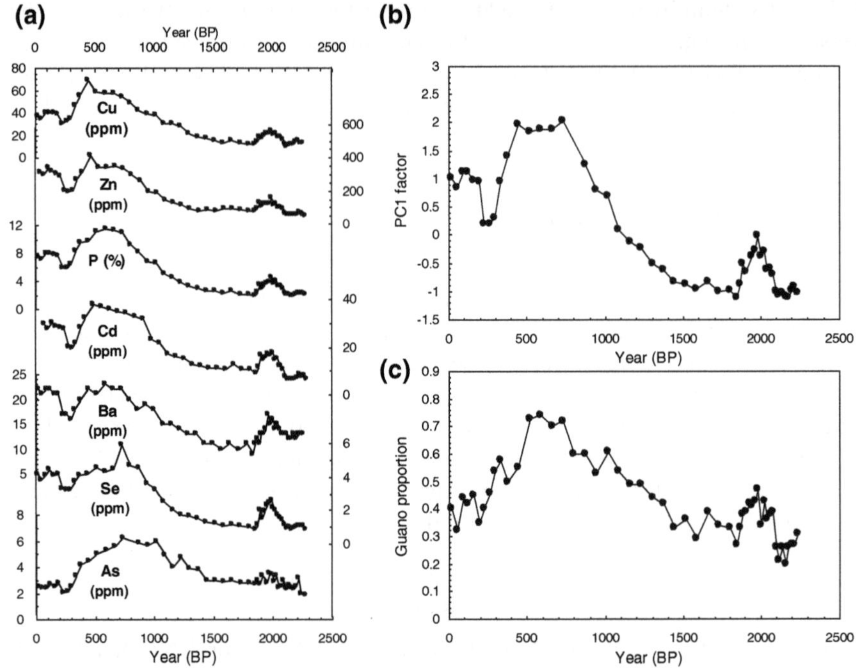


**Fig. 7.1** Comparison between seabird population records obtained from chemical analysis and reflectance spectroscopy. **a** Levels of Cu, Zn, P, Cd, Ba, Se and As in bulk sediments of GQ. **b** Seabird population record (PC1 factor) on Ganquan Island over the past 2200 years (The results of PC1 factor is after Xu et al. 2011). **c** Concentration-versus-depth profiles of guano proportion obtained from reflectance spectroscopy

statistical difference between seabird population obtained from reflectance spectroscopy (Fig. 7.1c) and from bio-elements (Fig. 7.1b) ($R = 0.82$, $p < 0.05$), with minor disagreements in the upper layers. This discrepancy in upper layers may be attributed to the development of vegetation on these islands (see Chap. 6). If the upper samples are removed in the statistics, we obtained an even better correlation ($R = 0.92$, $p < 0.01$).

From the temporal distributions of guano percentage in Fig. 7.1c, seabird on this island over the past 2200 years follows a "settlement—expansion—recession" mode. Seabirds began to inhabit Ganquan Island more than two millennia ago, and then their numbers reached a small peak during the period 2100–1850 BP. Seabird population began to decrease at approximately 2000 BP, but again their abundance began to increase gradually since 1850 BP and began to peak around 900 BP, remaining at a peak for approximately 600 years. Guano proportion reached ~75 % when seabirds were most abundant during the Little Ice Age. This is in parallel with our earlier conclusion: a cool climate seems more favorable to seabirds on the Xisha Islands.

### 7.3.2 Stable Carbon ($\delta^{13}C$) and Nitrogen Isotope ($\delta^{15}N$) Compositions in Collagen Samples from Ganquan Island

A broad survey of filed and laboratory data confirmed that there was a significant linear relationship between the stable nitrogen isotope values ($\delta^{15}N$) of an organism and its diet, and thus $\delta^{15}N$ is a robust indicator for organism's trophic position. Generally, an increase in one trophic level is accompanied by an increase in $\delta^{15}N$ values of approximately 3–5 ‰ due to trophic fractionation (Wade et al. 1991; Hobson 1999; Post 2002). Bird bones contain natural occurring protein collagen, which is rich in nitrogen. To track animal migration from stable nitrogen isotopes in ancient collagen samples, the quality of collagen samples is very important. Ambrose (1990) hypothesized that the concentration of carbon and nitrogen in well-preserved collagen was in the range from 15.3 % to 47 % and from 5.5 % to17.3 %, respectively. Elemental C/N was also applied to examine collagen quality, and collagen, with C/N values in the range of 2.9–3.6, was regarded as uncontaminated sample (DeNiro 1985). Elements C and N in the extracted collagen samples of GQ and modern bone collagen samples are listed in Table 7.1. C and N concentrations, as well as the C/N ratio, in all the collagen samples satisfied those criteria, so the collagen in our ancient bone samples was well-preserved and suitable for isotopic analysis.

Plots of collagen $\delta^{15}N$ and$\delta^{13}C$ against depth are given in Fig. 7.2a. To our knowledge, this is the first long-term (~2200 years) record of seabird diets from the South China Sea. According to diet-switch test, stable isotope signals in avian bone proteins show slower turnover rates than some other bird tissues, for example liver, muscle and blood (Hobson 1992). Generally, collagen $\delta^{15}N$ and $\delta^{13}C$ compositions of normal bone proteins of an organism offer a long time, in most cases a lifetime average, dietary intake. As a result, $\delta^{15}N$ and $\delta^{13}C$ values of collagen samples will have a high affinity with long-term patterns of habitat use of an organism and its food consumption (Pate 1998; Hu et al. 2009).

Typically, tissue $\delta^{13}C$ is frequently used to study source of primary producer and thus foraging areas of a predator. In contrast, $\delta^{15}N$ is a robust tool to investigate an organism's trophic position (France 1995; Post 2002). In an oceanic environment, inshore and benthic vegetation have heavier $\delta^{13}C$ than offshore and

**Table 7.1** Statistical analysis of C and N in the collagen samples from GQ profile

| Sample no | Item | Min | Max | Mean | Standard deviation |
|---|---|---|---|---|---|
| GQ (n = 15) | N | 11.40 % | 14.83 % | 13.87 % | 1.12 % |
| | C | 32.04 % | 42.07 % | 39.10 % | 3.03 % |
| | C/N(mol) | 3.21 | 3.41 | 3.29 | 0.05 |
| Modern (n = 7) | N | 14.34 % | 16.25 % | 15.48 % | 0.74 % |
| | C | 38.94 % | 44.26 % | 42.50 % | 1.90 % |
| | C/N(mol) | 3.14 | 3.25 | 3.20 | 0.05 |

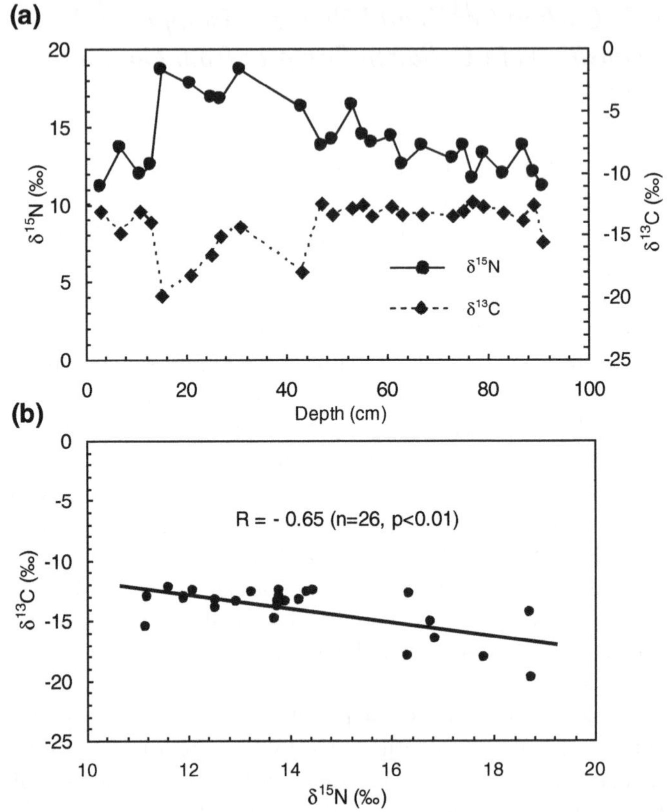


**Fig. 7.2** **a** Vertical distributions of $\delta^{15}N$ and $\delta^{13}C$ in collagen samples of GQ; **b** correlation between collagen $\delta^{15}N$ and $\delta^{13}C$ [Reprinted from Xu et al. (2014), Copyright (2014), with permission from Elsevier]

phytoplankton, respectively, implying that it is feasible to identify food source via stable carbon isotope analysis (France 1995; Forero and Hobson 2003). For our samples, the value of collagen $\delta^{13}C$ is $-14.12 \pm 1.97$ ‰ (n = 26). In terms of absolute value, such a level is relatively high. For instance, Chamberlain et al. (2005) analyzed stable carbon isotopes in protein of avian fossil bone samples from the middle-latitude California, and they reported quite low levels of $\delta^{13}C$ ($<-20$ ‰) in fossil samples. Another study of Antarctic penguin tissues reported even lower $\delta^{13}C$ values (Cherel et al. 2008). The elevated level of $\delta^{13}C$ in collagen samples from the Xisha Islands might be attributed to latitude effect of some isotopes. For example, $\delta^{13}C$ in organisms generally show decreasing trend along with latitude increase (Takai et al. 2000; Cherel et al. 2008).

From Fig. 7.2a, we observe that there is a minor change in both collagen $\delta^{15}N$ and $\delta^{13}C$ for the samples below the depth 43 cm in the core GQ. Collagen $\delta^{13}C$ ($-13.14 \pm 0.77$ ‰, n = 16) in such samples shows a quite small variance, implying that seabirds had a similar foraging behavior and their foraging strategy remains relatively stable. Given that collagen $\delta^{13}C$ remains at a high level, we suggest seabirds on the Ganquan Island probably preyed principally on inshore organisms as their main food. Normally, an increase in one trophic position is followed by an increase in $\delta^{15}N$ values of roughly 3–5 ‰ (Wade et al. 1991; Post 2002). Below 43 cm depth, average value of $\delta^{15}N$ in collagen samples is $13.38 \pm 1.29$ ‰ ($n = 16$), suggesting a very narrow ecological niche for seabirds. This is consistent with a diet based mainly on squid and flying fish, and the hypothesis that tropical seabirds overall occupy a very narrow isotopic niche, with significant overlaps among different bird species. This is explained by the fact that tropical ocean has a low productivity and birds have a similar foraging strategy (Cherel et al. 2008). In contrast to samples below 43 cm depth, both $\delta^{15}N$ and $\delta^{13}C$ in collagen samples at the depths from 43 to 15 cm show significant fluctuations (Fig. 7.2a), unveiling a substantial change in seabird foraging habitat and their trophic level. The $\delta^{13}C$ values in these samples change from $-19.79$ to $-14.33$ ‰. This suggests a possible broad foraging sites and a remote foraging distance from inshore to offshore areas. This may eventually render the carbon isotopes of bone collagen samples more negative. Likewise, vertical distribution of $\delta^{15}N$ at the same depths also shows evident change, varying at a large range of 7.63 ‰ (from 11.17 to 18.74 ‰). Given that tissue $\delta^{15}N$ increases by a approximately 3.4 ‰ for an increase in one trophic level (Post 2002), the seabirds on the Ganquan Island over the past two millennia encompassed almost two trophic positions. We suggest that foraging strategy and trophic levels of seabirds on the Ganquan Island have changed significantly over the past two millennia, and seabirds in tropical areas do not always occupy a narrow ecological niche. The substantial change of collagen $\delta^{15}N$ and $\delta^{13}C$ during the time from 43 to 15 cm depth of the profile GQ coincides with to a larger seabird population size (See Sect. 7.3.3 ). The recent decrease in $\delta^{15}N$ and increase in $\delta^{13}C$ are also reflections of seabird ecology and climate changes after the cool and dry Little Ice Age.

### *7.3.3 Changes of Seabird Diets in Response to Seabird Population Dynamics*

Intra/inter-specific competitions among seabirds are closely related to their population density, and this would affect seabirds' foraging strategy and further dietary structures. We hypothesize that there will be an inherent link between seabird population size and their trophic levels on the Ganquan Island. To test our tentative theory, we compared the record of reconstructed seabird population with isotope-based trophic level.

Distributions of collagen $\delta^{15}N$ and seabird population as a function of time in the GQ profile are plotted in Fig. 7.3. It is interesting that seabird population record from both bio-elements (PC1) and spectrum (guano proportion) changed with seabird trophic levels. A high level of seabird population, as well

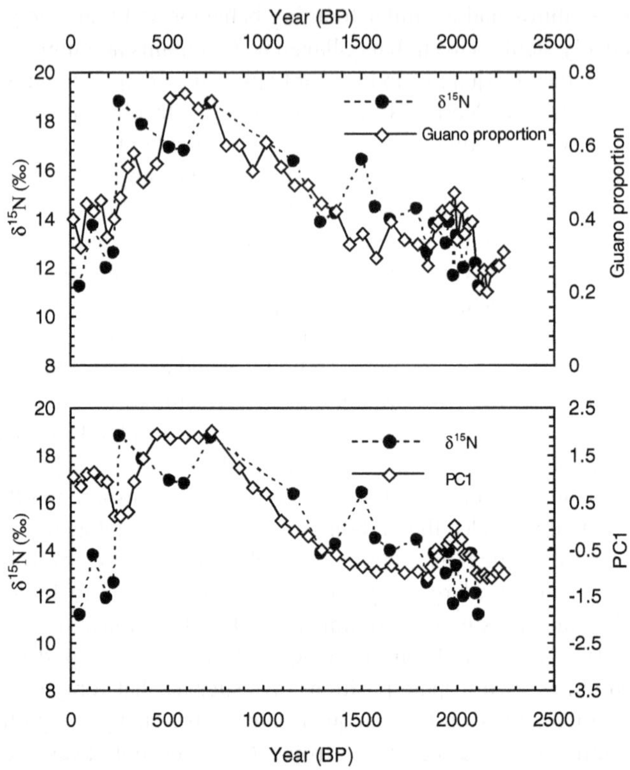


**Fig. 7.3** Comparison between seabird trophic level ($\delta^{15}N$) and seabird population records. Seabird population is shown as guano proportion and principle component one (PC1) deduced from bio-element concentrations [Reprinted from Xu et al. (2014), Copyright (2014), with permission from Elsevier]

as elevated collagen $\delta^{15}N$ (17.78 ± 0.95 ‰, n = 4), appeared during the time 1200–200 BP, which encompasses almost the entire cold Little Ice Age. By contrast, both seabird population size and collagen $\delta^{15}N$ are relatively lower than before (13.55 ± 1.44 ‰, n = 17) and after (12.34 ± 1.07 ‰, n = 4) that time. We believe the fluctuations of collagen $\delta^{15}N$ in the Ganquan Island were attributed to changes in seabird trophic positions. A number of studies have suggested that dietary changes within a seabird species, in response to either anthropogenic activity or oceanographic changes, were highly linked with their population density (Kitaysky et al. 2006; Trivelpiece et al. 2011; Wiley et al. 2013). Collagen $\delta^{15}N$ on the Ganquan Island principally depends on several factors: the foodweb baseline available to the seabirds, seabird species occupying different isotopic niches, and their dietary structures. As reported in previous research, Kienast (2000) found that $\delta^{15}N$ in organic matter from the South China Sea remained relatively stable over long time scales, suggesting an insignificant change in baseline of the maritime foodwebs. Seabird species changes might also result in alterations of collagen

$\delta^{15}N$. Although many seabird species sometimes occupy an individual island, they normally have similar dietary compositions and thus encompass similar isotopic niches (Cherel et al. 2008). A much earlier study analyzed 800 food samples from different seabird species on a tropical cay. It was also found that different bird species preyed on the same species, i.e. squid and flying fish (Ashmole and Ashmole 1967). From observation, frigatebirds, boobies and terns are the most common bird species on the Xisha Islands (Cao et al. 2007), and those seabirds show similar foraging behavior and almost exactly the same dietary structures (Balance and Pitman 1999). On a basis of similar trophic level, changes in bird species thus cannot explain substantial fluctuation in collagen $\delta^{15}N$. Consequently, we suggest that variability of collagen $\delta^{15}N$ in the Ganquan Island are probably ascribed to seabird dietary fluctuations.

According to previous study, squid and flying fish are two kinds of favorite food for tropical seabirds (Weimerskirch et al. 2008; Yin et al. 2008). Flying fish alone accounts for 70–93 % of the total diet for the birds on the Xisha archipelago, and squid and flying fish contribute approximately 95 % of birds' food, with only a minor part of unidentified prey species (Cao 2005). In subtropical and tropical regions, squid has a greater value of $\delta^{15}N$ than flying fish and thus occupies a higher trophic level (Cherel et al. 2008; Bugoni et al. 2010). Using museum samples, Norris et al. (2007) found that feather $\delta^{15}N$ (or trophic level) of marbled murrelet was closely correlated to fish proportion in the diet. For our study, we suggest that the 2000-year record of collagen $\delta^{15}N$ in our bone samples probably reflects the historic relative contributions of flying fish and squid to the diets of seabirds on the Ganquan Island.

It is probable that seabirds on the Ganquan Island ate more squid and the unidentified high-trophic species during the seabird thriving time 1200–200 BP, which caused elevated levels of collage $\delta^{15}N$ in avian bone tissues. In contrast, seabirds perhaps feeded on more low-trophic flying fish in the periods before 1200 BP and after 200 BP, ultimately rendering values of $\delta^{15}N$ in bone collagen samples more negative. As flying fish "flies" on the ocean surface, they are easy to be captured by seabirds. When there was a high level of seabird population during the period 1200–200 BP, more food was required to support those birds. Flying fish were thus in relatively short supply and these birds were forced to capture their alternative prey squid. The birds might also began to eat organisms that they seldom prey on, for example young *Xiphias gladius* ($\delta^{15}N = 14.05 \pm 0.69$ ‰), *Thunnus obesus* ($\delta^{15}N = 13.88 \pm 0.07$ ‰, Cherel et al. 2008). During seabird-thriving period, competitions among individual seabirds may render the them to dive deeper or travel to more remote areas where prey was more abundant or easier to catch. Except for the similar distributions of seabird trophic level and population size (Fig. 7.3), collagen $\delta^{15}N$ and $\delta^{13}C$ are significantly and negatively correlated (Fig. 7.2b). This further implies a change of seabirds' foraging behavior from inshore to offshore areas under fierce intra/inter-specific competitions. Researchers have investigated the spatial distributions of $\delta^{13}C$ and $\delta^{15}N$ in aquatic ecosystems: $\delta^{13}C$ shows a decreasing trend from inshore to offshore areas (Hobson 1999), while $\delta^{15}N$ in fish expresses an positive gradient along the same pathway.

For example, France (1995) found that $\delta^{15}N$ in marine organisms were more positive than those inhabiting freshwaters, and organisms in anadromous and estuarine areas had intermediate $\delta^{15}N$ levels. Lewis et al. (2001) reported that seabirds from larger colonies traveled further to seek food, and there was a positive correlation between mean foraging range and colony size. These findings support our theory that competitions for prey may lead to the changes of seabird foraging strategy. For example, seabirds may dive deeper and/or travel further to capture high-trophic organisms as their prey. Collagen $\delta^{15}N$, in combination with $\delta^{13}C$, revealed an unambiguous dietary change for seabirds on the Ganquan Island during the past two millennia. All the evidence supports our hypothesis that there had been a gradual change in seabirds' foraging range from coastal area to pelagic regions, when the seabird population began to increase significantly since approximately 1,200 BP. Using avian subfossils from the Hawaiian Islands and C/N isotope technology, Wiley et al. (2013) demonstrated a 3000-year record of seabird diet, and a significant influence of anthropogenic fishing activity on trophic position of birds was identified. In comparison with their finding, it seems that the Ganquan Island received a much lower level of human impact than the Hawaii Islands. For the future research, study of oceanic isoscape in the South China Sea will be helpful for a better interpretation of isotopic composition in avian subfossils.

## 7.4 Brief Summary

Well-preserved bone remains were collected from Ganquan Island. Seabird palaeodiet was investigated by stable carbon and nitrogen isotopes, and the results showed that seabird population was closely related to its trophic position: more seabirds corresponded to higher collagen $\delta^{15}N$, or trophic level. When there are more seabirds, a fierce intra/inter-specific competition may force birds to migrate from inshore habitats to offshore areas, and feed on preys with higher tropic level. The significantly negative correlation between collagen $\delta^{13}C$ and $\delta^{15}N$ further supports our hypothesis.

## References

Ambrose SH (1990) Preparation and characterization bone and tooth collagen for stable carbon and nitrogen isotope analysis. J Archaeol Sci 17:431–451

Ashmole NP, Ashmole MJ (1967) Comparative feeding ecology of sea birds of a tropical oceanic island. Peabody Museum of Natural History, Yale University, New Haven

Balance LP, Pitman RL (1999) Foraging ecology of tropical seabirds. In: Adams NJ, Slotow RH (eds) Proceedings of 22 international Ornithol Congr Durban, pp 2057–2071

Bearhop S, Adams C, Waldron S, Fuller RA, Macleod H (2004) Determining trophic niche width: a novel approach using stable analysis. J Anim Ecol 73:1007–1012

Brown TA, Nelson DE, Vogel JS, Southon JR (1988) Improved collagen extraction method by modified Longin method. Radiocarbon 30(2):171–177

Bugoni L, McGill RAR, Furness RW (2010) The importance of pelagic longline fishery discards for a seabird community determined through stable isotope analysis. J Exp Mar Biol Ecol 391:190–200

Cao L (2005) Population ecology of the red-footed booby on the Xisha Archipelago. In: PhD dissertation. Lanzhou University

Cao L, Pan YL, Liu NF (2005) Status of the red-footed booby on the Xisha Archipelago, South China Sea. Waterbirds 28(4):411–419

Cao L, Pan YL, Liu NF (2007) Waterbirds of the Xisha Archipelago, South China Sea. Waterbirds 30(2):296–300

Chamberlain CP, Waldbauer JR, Fox-Dobbs K, Newsome SD, Koch PL, Smith DR, Church ME, Chamberlain SD, Sorenson KJ, Risebrough R (2005) Pleistocene to recent dietary shifts in California condors. Proc Natl Acad Sci USA 102:16707–16711

Cherel Y, Hobson KA (2007) Geographical variation in carbon stable isotope signatures of marine predators: a tool to investigate their foraging areas in the Southern Ocean. Mar Ecol-Prog Ser 329:281–287

Cherel Y, Corre ML, Jaquemet S, Ménard F, Richard P, Weimerskirch H (2008) Resource partitioning within a tropical seabird community: new information from stable isotopes. Marine Ecol Prog Ser 366:281–291

DeNiro MJ (1985) Postmortem preservation and alteration of in vivo bone collagen isotope ratios in relation to palaeodietary reconstruction. Nature 317:806–809

Emslie SD, Patterson WP (2007) Abrupt recent shift in $\delta^{13}C$ and $\delta^{15}N$ values in Adelie penguin eggshell in Antarctica. Proc Natl Acad Sci USA 104:11666–11669

Exploration Group of Xisha Islands of Institute of Soil Science of Chinese Academy of Sciences (CAS) (1977) Soil and guano phosphorus mine in Xi-Sha. Science Press, Beijing. (in Chinese with English abstract)

Forero MG, Hobson KA (2003) Using stable isotopes of nitrogen and carbon to study seabird ecology: applications in the Mediterranean seabird community. Sci Mar 67:23–32

France R (1995) Stable nitrogen isotopes in fish: literature synthesis on the influence of ecotonal coupling. Estuar Coast Shelf Sci 41:737–742

Hilton GM, Thompson DR, Sagar PM, Cuthbert RJ, Cherel Y, Bury S (2006) A stable isotopic investigation into the causes of decline in a sub-Antarctic predator, the rockhopper penguin Eudyptes Chrysocome. Glob Change Biol 12:611–625

Hobson KA (1992) Assessing avian diets using stable isotopes I: turnover $^{13}C$ in tissues. Condor 94:181–188

Hobson KA (1999) Tracing origins and migration of wildlife using stable isotopes: a review. Oecologia 120:314–326

Hu YW, Shang H, Tong H, Nehlich O, Liu W, Zhao CH, Yu JC, Wang CS, Trinkaus E, Richards MP (2009) Stable isotope dietary analysis of the Tianyuan 1 early modern human. Proc Natl Acad Sci USA 106:10971–10974

Huang T, Sun LG, Stark J, Wang YH, Cheng ZQ, Yang QC, Sun S (2011) Relative changes in krill abundance inferred from Antarctic fur seal. PLoS ONE 6(11):e27331. doi:10.1037/journal.pone.0027331

Kienast M (2000) Unchanged nitrogen isotopic compositions of organic matter in the South China Sea during the last climatic cycle: global implications. Plaeoceanography 15:244–253

Kitaysky AS, Kitaiskaia EV, Piatt JF, Wingfield JC (2006) A mechanistic link between chick diet and decline in seabirds? Proc R Soc B 273:445–450

Lewis S, Sherratt TN, Hamer KC, Wanless S (2001) Evidence of intra-specific competition for food in a pelagic seabird. Nature 412:816–819

Longin R (1971) New method of collagen extraction for radiocarbon dating. Nature 230:241

Newsome SD, Clementz MT, Koch PL (2010) Using stable isotope biogeochemistry to study marine mammal ecology. Marine Mammal Sci 26(3):509–572

Norris DR, Arcese P, Preikshot D, Bertram DF, Kyser TK (2007) Diet reconstruction and historic population dynamics in a threatened seabird. J Appl Ecol 44:875–884

Pate FD (1998) Bone collagen stable nitrogen and carbon isotopes as indicators of past human diet and landscape use in southeastern South Australia. Aust Archaeol 46:23–29

Post DM (2002) Using stable isotopes to estimate trophic position: models, methods and assumptions. Ecology 83(3):703–718

Takai N, Onaka S, Ikeda Y, Yatsu A, Kidokoro H, Sakamoto W (2000) Geographical variations in carbon and nitrogen stable isotope ratios in squid. J Mar Biol Ass UK 80:675–684

Trivelpiece WZ, Hinke JT, Miller AK, Reiss CS, Trivelpiece SG (2011) Variability in krill biomass links harvesting and climate warming to penguin population changes in Antarctica. Proc Natl Acad Sci USA 108(18):7625–7628

Wade E, Mizutani H, Minagawa M (1991) The use of stable isotopes for food web analysis. Crit Rev Food Sci Nutr 30(4):361–371

Weimerskirch H, Corre ML, Bost CA (2008) Foraging strategy of masked boobies from the largest colony in the world: relationship to environmental conditions and fisheries. Mar Ecol Prog Ser 362:291–302

Wiley AE, Ostrom PH, Welch AJ, Fleischer RC, Gandhi H, Southon JR, Stafford TW, Penniman JF, Hu D, Duvall FP, James H (2013) Millennial-scale isotope records from a wide-ranging predator show evidence of recent human impact to oceanic food webs. Proc Natl Acad Sci USA 110(22):8972–8977

Xu LQ, Liu XD, Sun LG, Yan H, Liu Y, Luo YH, Huang J (2011) A 2200-year record of seabird population on Ganquan Island, South China Sea. Acta Geol Sin-Engl 85:957–967

Xu LQ, Liu XD, Jiang S (2014) Late-holocene seabird palaeodietary record from Ganquan Island, South China Sea. Quat Nat Int 333:139–145

Yin XB, Xia LJ, Sun LG, Luo HH, Wang YH (2008) Animal excrement: a potential biomonitor of heavy metal contamination in the marine environment. Sci Total Environ 399:179–185

# Chapter 8
# Evidence of Human Activities from the Ornithogenic Sediments of the Xisha Islands

Human activity is a factor that impose significant impacts to the natural environment, even in some remote areas, since onset of the Holocene. In this chapter, we will use proxy data to examine possible influences of anthropogenic activities on the Xisha Islands. Proxies that we use include heavy metal mercury (Hg) and black carbon (BC), which are closely related to human civilization.

## 8.1 A 700-Year Record of Mercury in Avian Eggshells of Guangjin Island, South China Sea

### *8.1.1 Analytical Method*

Mercury is a ubiquitous heavy metal element with high toxicity in the environmental. It can be enriched in biological materials due to biomagnification, and thus has drawn increasing attention in recent years. We attempt to analyze its levels in avian tissues, as well as the bulk sediments.

A large number of eggshell fragments were collected from the duplicate sediment profile GJ3. Prior to analysis, the eggshell debris were processed as follows. The samples were placed in a beaker and washed by distilled water under an ultrasonic environment to remove clay, faeces, etc. The washed eggshells were dried at a temperature of 60 °C for 12 h; approximately 0.1 g subsamples were taken and precisely weighed. Eggshell fragments were then crushed and digested by high purity grade $HNO_3$ plus $H_2O_2$ to a transparent solution (Sun et al. 2006). The concentrations of total Hg were determined by the instrument Atomic Fluorescence Spectrophotometer (Manufactured by Beijin Jitian, AFS-930). For quality assurance (QA) and quality control (QC), standard reference material GBW07108 was used in each batch of analysis, and the determined value from AFS analyses were in agreement with the reference range. All Hg concentrations in eggshells were expressed on a dry-weight basis.

 
L. Xu, *Impact of Climate Change and Human Activity on the Eco-environment*, Springer Theses, DOI 10.1007/978-3-662-45003-1_8

## 8.1.2 Levels of Hg in Eggshells from Guangjin Island of the Xisha Archipelago

(1) Ultrastructural analysis of eggshells by Scanning Electron Microscope (SEM)

To examine whether these subfossil eggshells have experienced significant diagenetic alterations, we analyzed ultrastructure of both modern and ancient eggshells by scanning electron microscope (SEM). The avian eggshell is mainly composed of calcium carbonate (95 %), which is in the form of calcite. The calcite acts as skeleton of the shell, in which some organic constituents are embedded. In addition to eggshell membrane, several layers are observed from outside to inside of an eggshell: outer layer, spongy layer and mammillary cone layer (Ma et al. 2005). The SEM graphs of ancient eggshells, as well as fresh ones, are given in Fig. 8.1.

Mammillary cone layer is the most inner part of an avian eggshell, and its thickness is approximately 1/3 of the shell. Normally, it appears in the form of cone-shaped polygon, on which shell membrane fibers anchor (Nys and Gautron 2007). Mammillary cone layers are petal-like, and such a feature was clearly observed in the micrographs from fresh and subfossil eggshells (A-1, F-1 in Fig. 8.1) and they are very similar to each other in both shape and structure. The papillae of the inner surface are petal-shaped and there are large gaps among the papillae, including pore openings on the inside surface.

Spongy layer, also known as column layer or surface crystal layer, is also roughly 1/3 of the total shell thickness. The SEM photo of modern eggshell evidently showed the "sponge" in this layer (Fig. 8.1). It was observed that the

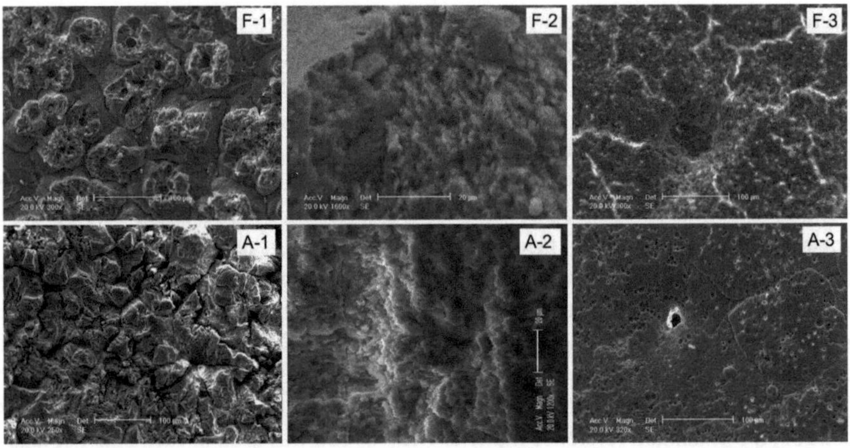


**Fig. 8.1** Comparisons between fresh and ancient eggshell SEM photos. *Note F-1–F-3* and *A-1–A-3* are the SEM photos of fresh (*F*) and ancient (*A*) eggshells respectively. From *left* to *right* Mammillary cone layer, spongy layer and outer layer of the shells [Reprinted from Xu et al. (2011b), Copyright (2001), with permission from Elsevier]

micrographs of spongy layer from ancient samples (F-2 in Fig. 8.1) were almost exactly the same as that of modern ones (A-2, in Fig. 8.1).

The outermost part of an eggshell is called outer layer, which has a relatively dense structure. The regular granular spherical crystals of the outer layer are observed in SEM graphs from both ancient and fresh eggshells. These particles can easily change to aragonite or calcite in aqueous solution. As a result of a relatively high temperature, the Xisha Islands is perennially dry. This leads to a poorly developed groundwater system on the sampling sites of the islands, and thus the coral sand-dominated sediments have a very low level of water content. All of these conditions are favorable for preservation of bird remains, including eggshell. Moreover, the irregular tortoise cracks and eggshell pores, which allow gaseous exchange, can be readily identified from these photos (Fig. 8.1, F-3, A-3). Despite of the some minor differences, outer layer structure of ancient eggshells resemble that of modern ones.

Based on the SEM analyses of the eggshell samples, it is found that the spongy layer and mammillary cone layer of ancient eggshells are quite similar to the respective layers of fresh *Sula sula* eggshells. In terms of morphology, though there is minor difference for the outer layer, both ancient and fresh eggshells had a similar outer layer structure. Thus, the ultrastructure of eggshell fragments is pretty similar to those of fresh ones.The diagenetic is insignificant, although those samples are hundreds of years old.

(2) The record of eggshell Hg over the past 700 years

Plots of eggshell Hg as a function of time for the sediment core GJ3 are supplied in Fig. 8.2. The time-series record is partly incomplete due to a lack of seabird remains in some of the top sediments. It is evident that the Hg concentrations showed substantial fluctuations over the past seven centuries. In general, the eggshell Hg showed an increasing trend over the last 700 years, and had a abrupt increase since the onset of 19th century.

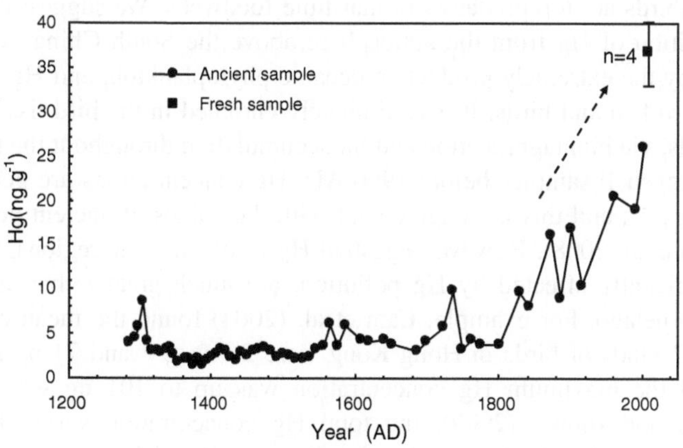


**Fig. 8.2** The 700-year record of eggshell Hg (based on dry mass) on Guangjin Island [Reprinted from Xu et al. (2011b), Copyright (2001), with permission from Elsevier]

The level of Hg in eggshell samples during the pre-industrial time from 1300 AD to 1800 AD remains at a low level with an average of 3.45 ng $g^{-1}$ (n = 53). After 1800 AD, the concentration of eggshell Hg increased rapidly, and its mean value during the period between 1800 AD and 2003 AD is 15.1 ng $g^{-1}$ (n = 9), which is more than 4 times greater than the pre-industrial times (i.e. before 1800 AD). In comparison with ancient eggshell samples, the average Hg concentration in modern eggshells (collected in 2008) is much higher (mean 36.7 ng $g^{-1}$, n = 4), and the maximum even exceeds 40 ng $g^{-1}$. The mean value is approximately 10 and 2.5 times higher than the period 1800–2003 AD and those of 1300–1800 AD, respectively. It is also observed that, of the 8 eggshell samples collected from the top 10 cm layers (i.e. recent ~100 years), the sample size of 4 samples was too small to be analyzed. The weights of the other 4 samples were also smaller than that below the 10 cm depth. The missing of recent eggshell fragments in the ornithogenic sediments suggests a decrease in the population size of seabirds over the last century possibly due to recent anthropogenic activities. The seabird habitat surroundings were significantly disrupted by some kinds of human activities, for example deforestation, guano-mining, house building, human-induced exotic species invasion (dog, cat etc.) (Zhao 1996). Integrity and independence of the fragile island ecosystem were substantially modified and disturbed (Cao 2005; Sun et al. 2005), perhaps ultimately leading to the abrupt decline of seabird population since the early 20th century.

According to previous studies of metals in bird eggs, the level of Hg in eggshells was pretty low, its concentrations in other avian tissues (e.g. albumen, feather and hair), however, were much higher. Albumen had a Hg concentration of approximately 220 ng $g^{-1}$, evidently higher than that of eggshell, while Hg in hair or feather could reach a level as high as thousands of ng $g^{-1}$ (Kennamer et al. 2005; Dietz et al. 2009). Average concentration of Hg in seawater is approximately 0.3 ng $ml^{-1}$ (Dai 1997); in contrast, the minimum content of eggshell Hg in the profile GJ3 is 3.45 ng $g^{-1}$, which is many times greater than its abundance in the ocean. Seabirds are top predators in maritime foodwebs. We suggest that the dry/wet deposition of Hg from the atmosphere above the South China Sea could be absorbed by the extremely productive oceanic phytoplankton, and Hg was further transfered to fish and birds. It was ultimately enriched in the bird tissues, as well as eggshells, via biomagnification and bioaccumulation throughout the food chain.

In the eggshell samples before 1900 AD, Hg concentrations are generally less than 10 ng $g^{-1}$, and this is in agreement with the values of ancient penguin eggshells (Yin et al. 2008). However, eggshell Hg levels in some regions, which have been significantly affected by Hg pollution, are much greater than those of the Xisha archipelago. For example, Lam et al. (2004) found the mean eggshell Hg levels for 2 kinds of birds in Hong Kong were 56 ng $g^{-1}$ and 71 ng $g^{-1}$ respectively, and the maximum Hg concentration was up to 101 ng $g^{-1}$. According to Pietrelli and Biondi (2009), the total Hg concentration varied from 26 to 200 ng $g^{-1}$ in Italian bird eggshells. As shown in Fig. 8.2, the eggshell Hg concentrations prior to 1800 AD in the Xisha Islands were relatively low, suggesting the minor influence of human activities on seabirds in ancient times. However, the

Hg levels of eggshells displayed rapid increase since 1800 AD, and the modern seabird eggshells had Hg concentration up to 36.7 ng $g^{-1}$ in the Xisha archipelago, approximate to the concentration in the eggshell samples taken from seriously contaminated area, suggesting that the anthropogenic pollutant might begin to threaten the survival of seabirds on the remote Xisha Islands. Indeed, the geochemical data on dissolved major, minor and trace constituents in groundwater and sediment samples from the Xisha coral islands showed that the present groundwater and sediments were seriously contaminated by pollutants including mercury, silver, cadmium etc. from mainland China and southeast Asia (Xie et al. 2005; Liu et al. 2008).

### *8.1.3 Environmental Implications of Hg in the Eggshells*

We believe that the eggshells in the profile GJ3 were well preserved after deposition. There is considerable evidence to testify to this: (a) Components of the ornithogenic sediments are relatively simple and they consist predominantly of coral sands, guano and plant humus. The eggshells in the GJ3 sediment profile were relatively young in age, and they were physically well preserved, as a result of a dry environment and the weak changes of physico-chemical conditions throughout the sediment profile, as well as very weak soil-forming processes. Additionally, field observation showed no significant visible differences among the ancient eggshell samples. (b) Hg has a low mobility (Lockhart et al. 2000) and a low exchange capacity, thus it is difficult to permeate into the interior of eggshells. (c) Scanning electron microscopy (SEM) analysis revealed that no significant diagenetic changes in the eggshells (refer to Sect. 8.1.2 and Fig. 8.1). Therefore, the weak post-alteration did not significantly change eggshell structure after deposition and it was reasonable to believe that the original Hg level, when the eggshells were formed, was well preserved.

Seabirds, as top predator of the marine food chain, accumulate Hg due to the effect of biomagnification. The composition of seabirds' diet could have a significant influence on their trophic level and the accumulation of Hg in eggshells. A broad survey of field and laboratory data confirms that there is a significant linear relationship between the stable nitrogen isotope values ($\delta^{15}N$) of an organism and its diet (Wade et al. 1991; Emslie and Patterson 2007). On this basis, the $\delta^{15}N$ can be used as a powerful proxy to infer trophic positions. Generally, an increase in one trophic level is accompanied by an increase in $\delta^{15}N$ values of approximately 3–5 ‰ (Wade et al. 1991; Post 2002). Eggshell formation, when initiated, is completed within a 24-h period, thus recording signs of the bird's most recent breeding diet. If the change of eggshell Hg concentration was critically subject to trophic transfer, we expected reasonably that the eggshell Hg content would be significantly correlated with $\delta^{15}N$ values of eggshell organic matter. The results of stable nitrogen isotope analysis showed that the $\delta^{15}N$ values of organic components in the ancient eggshell samples were $10.45 \pm 0.78$ ‰ ($n = 35$), and these values

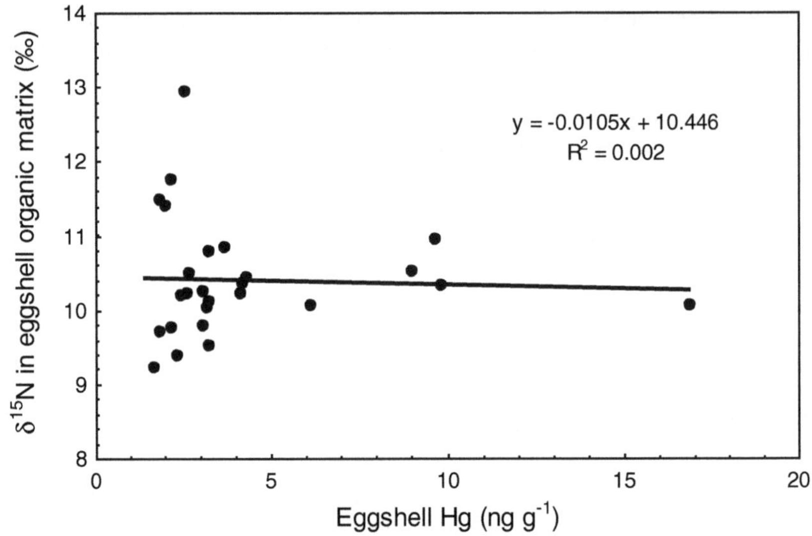


**Fig. 8.3** Correlation between eggshell Hg and $\delta^{15}N$ in eggshell organic matrix [Reprinted from Xu et al. (2011b), Copyright (2001), with permission from Elsevier]

have no statistical difference with the $\delta^{15}N$ values of modern egg membrane samples (10.11 ± 0.55 ‰, n = 4). The linear regression analysis showed that there was no significant correlation between the Hg concentrations and the $\delta^{15}N$ values of organic matrix in the eggshells ($R^2 = 0.002$, $P > 0.01$, Fig. 8.3), suggesting that the seabird trophic level change was not the main controlling factor for the change of eggshell Hg levels over the past 700 years. In short, here we suggested that both the post-depositional diagenetic alteration and trophic level changes of seabirds could not exert significant influences on the historical change of eggshell Hg concentrations. Thus, the Hg content in the eggshells was likely a marker for atmospheric Hg deposition, and the change of Hg levels recorded in the ancient eggshells in Guangjin Island likely reflected the past Hg pollution history in the marine environment.

### *8.1.4 Characteristics of Hg Record in the Eggshells Over the Past 700 Years*

The change of eggshell Hg over the past 700 years and a collection of Hg records from different places of the world are plotted in Fig. 8.4. In general, the Hg record of GJ3 has notable global distribution characteristics and also regional differences.

In the eggshell samples prior to 1550 AD, Hg content is relatively low, no more than 6 ng $g^{-1}$ with one anomaly 8.7 ng $g^{-1}$ at about 1300 AD and one pos-

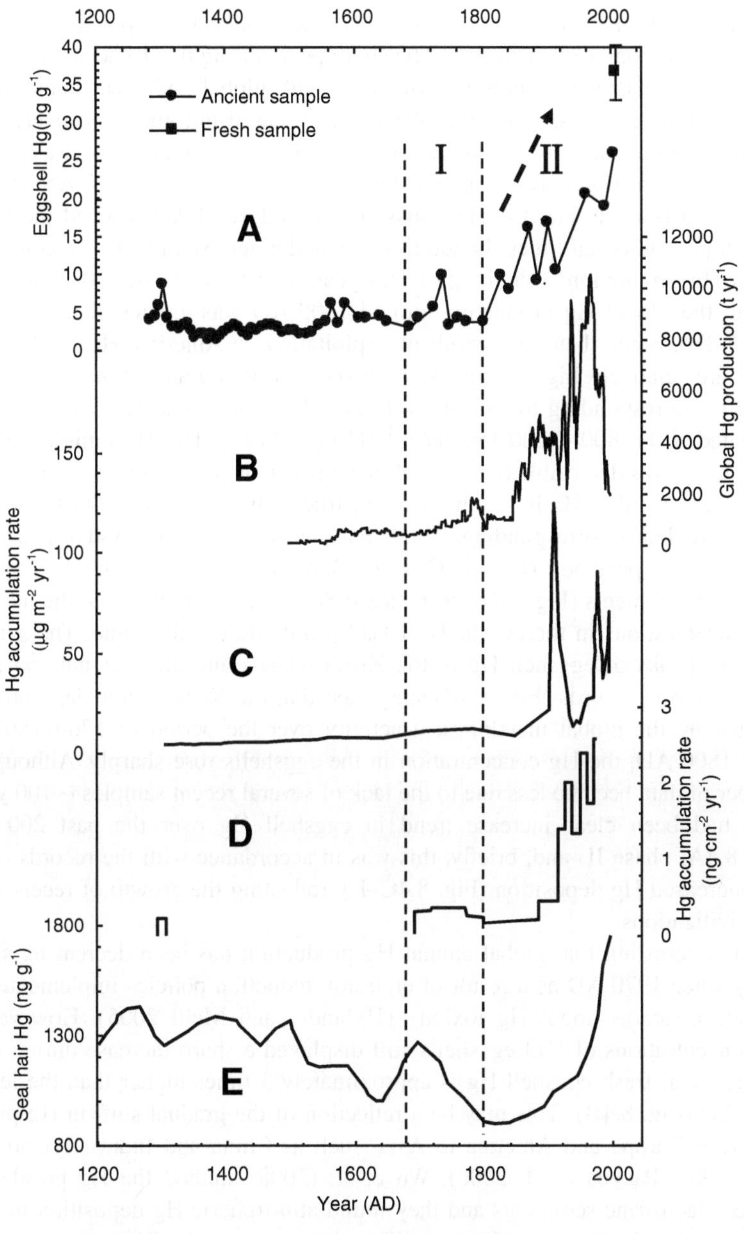


**Fig. 8.4** Comparisons among eggshell Hg on Guangjin Island and a group of reconstructed Hg records. Cited literature: *A* this study; *B* global Hg production over the past 500 years (Hylander and Meili 2003); *C* Hg accumulation rate over the past ~700 years from an peat bog in Swiss (Roos-Barraclough et al. 2002); *D* reconstructed Hg accumulation rate from varved lake sediments in Swedon (Renberg 1986); *E* A 2,000-year record of Hg in seal hairs from west Antarctica (Sun et al. 2006, based on dry mass). [Reprinted from Xu et al. (2011b), Copyright (2001), with permission from Elsevier]

sible explanation for the anomaly might be the sudden Hg deposition caused by volcanic eruption or other factors. The small peak during the decades around 1600 AD (Fig. 8.4A) may be closely associated with global and local human industrial activities. A massive amount of Hg was released into the atmosphere due to extensive mining activity in Europe and this large scale Hg emission was reflected in Spanish peat bog core samples (Martínez-Cortizas et al. 1999). According to their reconstructed data, the Hg emissions around 1600 AD reached the highest level in pre-industrial times. In addition, Hylander and Meili (2003) reconstructed global Hg production over the past 500 years and they discovered a similar Hg record: the global Hg production around 1600 AD was greater than ever before (Fig. 8.4B). Since then, as a result of exploitation on America, Hg levels in eggshells are relatively high and peaked between 1700 AD and 1800 AD (Fig. 8.4, phase I), corresponding to the relatively high level of annual Hg loss, which was estimated to be 800–1,200 ton year$^{-1}$ (Nriagu 1994). The Hg emission was so great that it was detectable by many global environmental archives. For instance, a 2000-year record of Hg in seal hairs reconstructed by Sun et al. (2006) unambiguously revealed a correspondingly high Hg consumption (Fig. 8.4E), and results from a Swiss peat bog (Fig. 8.4C, Roos-Barraclough et al. 2002) and Swedish lacustrine sediments (Fig. 8.4D, Renberg 1986) were in parallel with the increased Hg concentrations in the eggshells of GJ3 profile during that time. The synchronous Hg peaks of eggshell Hg in the Xisha Islands and the regional and global Hg production records (Fig. 8.4B) suggested that the Xisha archipelago might be affected by the global metallurgical activity over the period of 1700–1800 AD. After 1800 AD, the Hg concentration in the eggshells rose sharply. Although the number of data became less due to the lack of several recent samples (~100 years), there had been clear increase trend in eggshell Hg over the past 200 years (Fig. 8.4A, phase II) and, briefly, this was in accordance with the records of rapidly increased Hg deposition (Fig. 8.4C–E), reflecting the growth of recent industrial civilizations.

It is interesting that global annual Hg production has been decreasing significantly since 1970 AD as a result of emission reduction policies implemented due to wide concerns about Hg toxicity (Hylander and Meili 2005). However, the Hg concentrations of GJ3 eggshells still displayed a sharp increase during recent decades, e.g. fresh eggshell Hg is approximately 3 times higher than the level of 1850 AD (Fig. 8.4B). This may be a reflection of the gradual shift in Hg production from Europe and America to Asia, such as China and India (Hylander and Meili 2003, Pacyna et al. 2006). Wu et al. (2008) studied the Hg profile from Chinese lacustrine sediments and they found atmospheric Hg deposition in China did increase over the past 100 years. This demonstrates that Hg deposition in the Xisha Islands reflects global and regional distribution patterns. Therefore, the increase in Hg concentration of ancient eggshells in Guangjin Island of the Xisha archipelago before 1970 is likely to have been derived from global industrial sources. However, the rapid increase from 1970 AD to present is mainly attributed to the anthropogenic emissions of Asia due to the increase in coal combustion in the region related to rapid economic development since the 1970s (Jiang et al.

2006; Hong et al. 2009; Yang et al. 2010a). Over the past two decades anthropogenic Hg emissions have declined in North America and Europe but increased substantially in East Asia and India (Pacyna et al. 2006). Asia has become the largest contributor of anthropogenic atmospheric Hg, responsible for over half of the global emission in 2000 (Pacyna et al. 2006; Li et al. 2009). Selin et al. (2008) estimated anthropogenic enrichment factors for Hg deposition from preindustrial to present day in different parts of the world, and found that the highest enrichment factors are in anthropogenic source regions, exceeding 10 in East Asia.

According to the analysis of eggshell Hg spanning the past 700 years, we draw the conclusions below.

The Hg concentration in ancient eggshells was at least one order of magnitude larger than that of seawater, testifying to the biomagnification of Hg in marine ecosystems. There was little evidence for the suggestion that the fluctuation of Hg concentration in ancient eggshells could be due to post-depositional diagenetic changes or trophic level transfer of seabirds. Therefore, Hg in the eggshells was likely a marker for past atmospheric Hg deposition in marine environment. Eggshell Hg levels had tended to increase over the past 700 years, in particular over the past 200 years. Industrially related Hg inputs to the South China Sea and the Xisha Islands may be the most likely explanation for the increase. Mercury emissions from developed countries have been dramatically reduced since the 1970s, leading to reduced atmospheric Hg deposition; however, the Hg level in the eggshells was still increasing rapidly after 1970 AD. These eggshell data suggested that most of the increase in the Xisha islands during the last 30 years of the 20th century was primarily as a result of local, not global, deposition, implying that the Hg production center had gradually shifted from Europe and America to Asia. Historically, the Hg levels in fresh eggshell samples are far higher than those of ancient eggshells and still follow an evident upward trend, and thus in the future we should assess the potential seabird ecological risk in the Xisha Islands for the increasing mercury burden in eggs.

## 8.2 Historical Change of Mercury Pollution on the Remote Yongle Archipelago, South China Sea

### *8.2.1 Distributions of Hg Concentrations in the Ornithogenic Sediment Profiles and Its Depositional Fluxes*

We also investigate Hg in the bulk sediments from Guangjin, Jinqing and Jinyin Islands. The determination of Hg in the ornithogenic sediments of GJ3, JQ and JY2 are the same as that of eggshell Hg (For details, refer to Sect. 8.1.1). The concentrations of Hg versus depth in the three sediment cores showed quite a similar distribution pattern (Fig. 8.5). The bottom samples were featured by high levels of coral sand, low contents of organics and a very low concentration of Hg. For

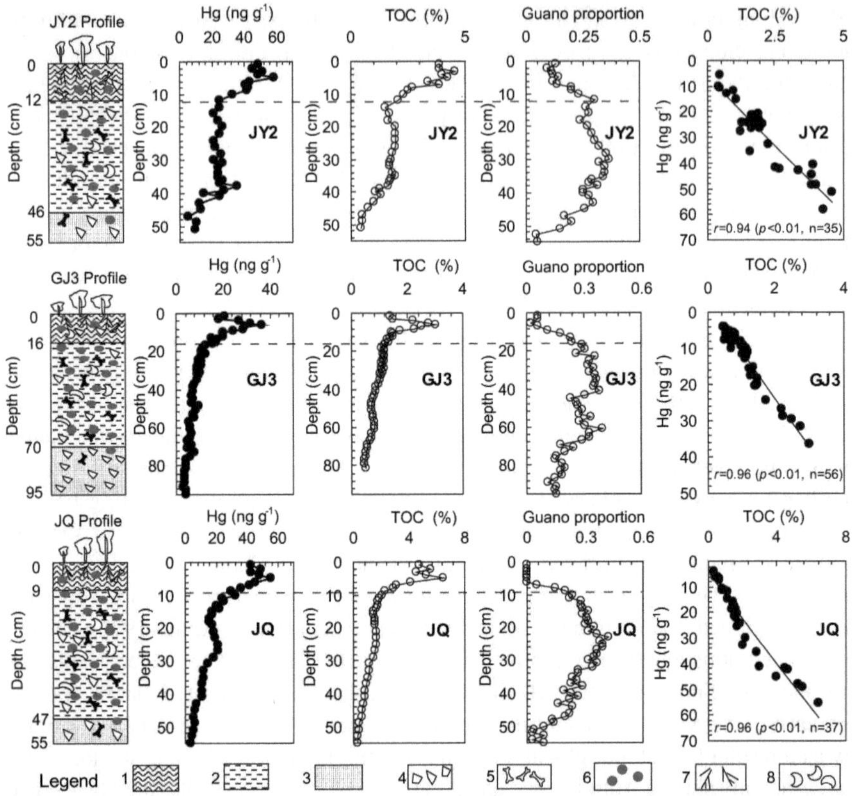

**Fig. 8.5** Vertical distributions of TOC, Hg and guano contents in the profiles GJ3, JQ and JY2 and correlations between TOC and Hg. Legend: *1* grey black humus, plant residues and medium to large size coral sands and a little bit guano and seabird bone remains in the upper layer; *2* light brown ornithogenic sediments containing a lot of medium grained coral sands, guano particles, bone remains and fish scales; *3* yellow to white coral sand sediment layer with low organic matter content and a few guano pellets and numerous large size calcareous bioclasts; *4* calcareous bioclasts; *5* bird and fish bones; *6* guano; *7* remains of plant leaf, stem and root; *8* fish scales. [Reprinted from Liu et al. (2012), Copyright (2012), with permission from Elsevier]

the three sediment cores, the levels of Hg in the bulk sediments below the critical depths of 12 cm (GJ3), 14 cm (JQ) and 10 cm (JY2) are generally less than 20, 20, 30 ng $g^{-1}$, respectively. The Hg levels in each of the sediment profiles increased with organic matter toward the surface, and peaked at the subsurface layer, but showed a slight decrease in the uppermost sediments. For example, the Hg profiles of GJ3, JQ and JY2 show evident peak values of 36.4, 55.1 and 58.1 ng $g^{-1}$ at the respective depths of 6, 5 and 5 cm.

Hg could be readily absorbed by organic matter, and becomes more enriched as the level of organic matter increases in soils or sediments (Yang and Rose 2003; Gil et al. 2010; Szopka et al. 2011). The Pearson correlation analysis between

TOC and Hg in the sediments of the three profiles yields significantly positive correlations with correlation coefficients better than 0.9 ($p < 0.01$) (Fig. 8.5), suggesting that the Hg accumulation in the three islands of the Yongle Islands depended mainly on sedimentary organics in the bulk sediments. A great number of research has shown that Hg is closely linked with organic matter input (Ravichandran 2004; Sanei et al. 2010; Stern et al. 2009; Feyte et al. 2010). As most Hg in the three profiles was bond to sediment organics, and the input of organic matter remains at a fairly stable level, it is suggested that down-ward migration of Hg, i.e. delivery to lower strata by rainfall, is insignificant. Sediment Hg contents have decreased in the uppermost sediments of the three profiles (Fig. 8.5), and this can be explained by the increase in sediment load, which diluted the Hg in the sediments.

According to our analysis in Chap. 5, organic matter in the bulk sediments below a specific depth of each profile is primarily sourced from bird droppings, but its increase in top layers is caused by the substantial increase in plant humus input in recent times (refer to Chap. 5 and Xu et al. 2011a). Following the method of Chap. 6 and Liu et al. (2011), vertical distributions of guano ratio in each of the three cores have been reconstructed using spectrum technique on a three-component basis (i.e. coral sand, guano and plant humus). As given in Fig. 8.5, the guano levels overall show similar distribution patterns with Hg contents below the critical depths (approximately 1850 AD) of the three profiles, and they were significantly and positively correlated (Fig. 8.6). This suggests that heavy metal Hg in the bulk sediments below the critical depths probably stemmed from guano, and this may generate a greater correlation coefficient between Hg and organic matter (Fig. 8.5). However, the level of guano proportion decreased significantly in the top layer of each sediment core. In contrast, Hg concentration had a substantial increase (Fig. 8.5), indicating that the bird dropping was no longer the main source of Hg in the top sediments. The abrupt increase in Hg in these upper sediments can be best explained by human-derived Hg emission. After Hg was released to the atmospheric in recent times, it was transferred to the Xisha Islands via atmospheric circulation, and formed chemically stable coordination complexes with the plant-derived organics, and eventually deposited at the sampling sites. Indeed, as reviewed by Fitzgerald et al. (1998), the atmospheric Hg deposition directly to soils, complexation of Hg by organic matter, and atmospheric inputs to leaves and hence to surface soils are well-documented processes in remote areas.

To eliminate the possible effects of source material inputs and sedimentation rate changes on Hg distribution, we used Hg deposition flux, rather than concentration, to assess time-series record of the Hg pollution (Kamman and Engstrom 2002; Yang and Rose 2003; Lindeberg et al. 2007). The results were given in Fig. 8.7. The deposition fluxes of Hg in the Jinqing Island, Guangjin Island and Jinyin Island of the Yongle archipelago remained at a relatively low level before 1850 AD (Fig. 8.7), implying a low natural background. The consistency of Hg flux records among those individual islands provides evidence that Hg accumulation in these profiles is not significantly affected by local geological sources or post-depositional diagenetic processes. Typically, Hg accumulation rate depended on both

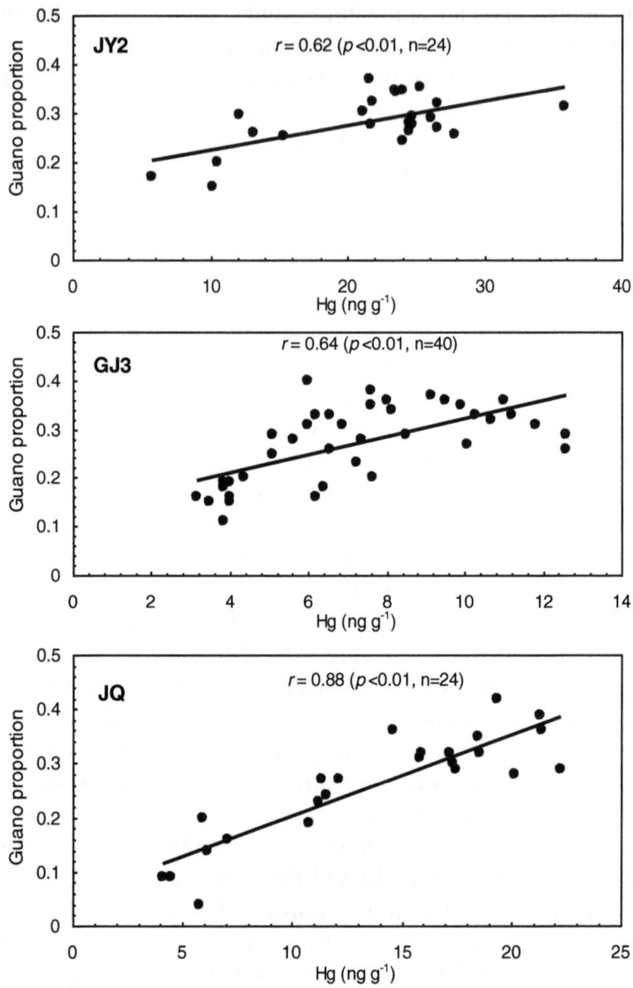


**Fig. 8.6** Correlations between guano proportion and Hg in the sediments of JY2, GJ3 and JQ below the critical depths [Reprinted from Liu et al. (2012), Copyright (2012), with permission from Elsevier]

direct atmospheric deposition and organics-associated Hg input. Many studies have found that the input of Hg to lacustrine sediments was determined by both autochthonous algae and surface runoff of terrigenous organics. Thus, the deposition flux of atmospheric Hg may be overestimated due to organic matter input (Phillips et al. 2011; Sanei et al. 2010; Outridge et al. 2007). As stated above, human-derived Hg emission is a feasible source of Hg in recent sediments. Meanwhile, Hg concentrations and organic matter yield a significant positive correlation, implying a potential influence of organic matter on the accumulation rate of Hg.

In order to eliminate the potential impacts of natural and guano inputs on the accumulation rate of Hg, adjusted anthropogenic Hg flux ($Hg_A$) was adopted to

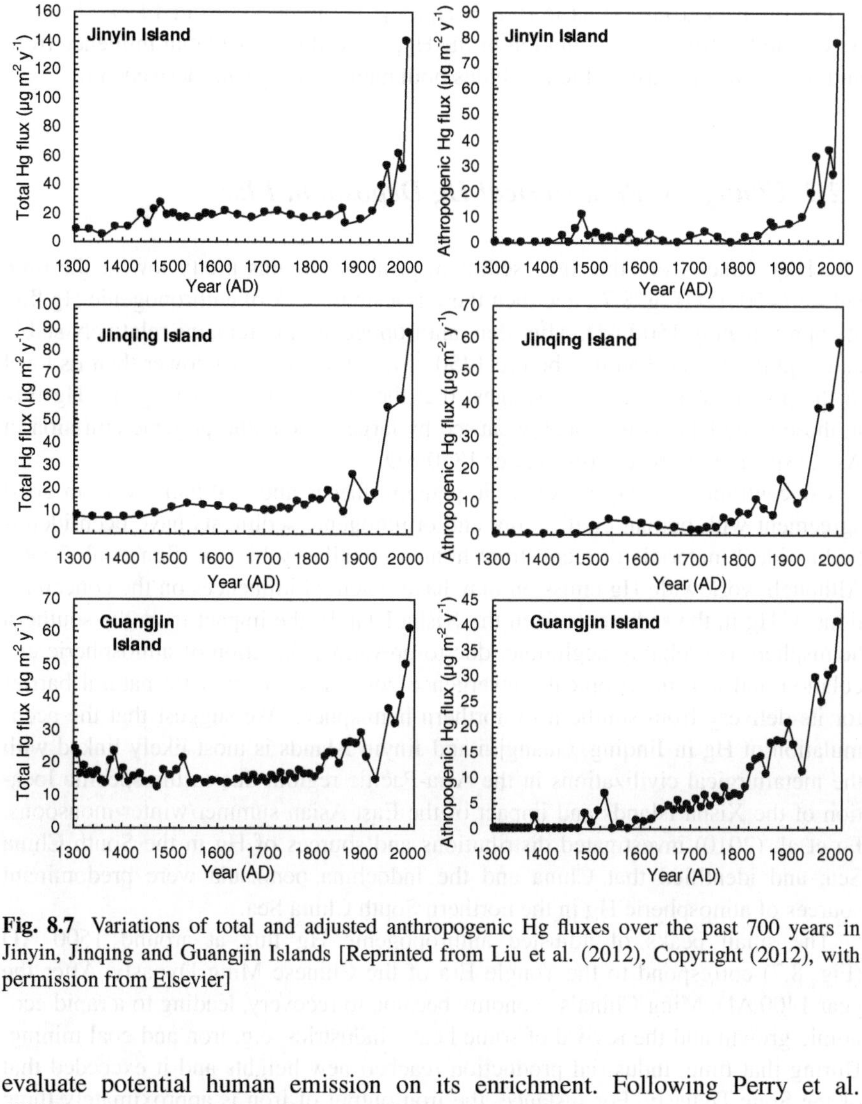


**Fig. 8.7** Variations of total and adjusted anthropogenic Hg fluxes over the past 700 years in Jinyin, Jinqing and Guangjin Islands [Reprinted from Liu et al. (2012), Copyright (2012), with permission from Elsevier]

evaluate potential human emission on its enrichment. Following Perry et al. (2005), anthropogenic Hg flux was calculated as follows:

$$Hg_A = [Hg_T - Hg_B - (Hg_B \times SR - Hg_B)]/FF$$

where $Hg_A$, $Hg_T$, and $Hg_B$ represent the deposition flux of adjusted anthropogenic Hg, total Hg, and background Hg in the sediments. SR (sedimentation ratio) in the equation is the ratio of post-industrial sedimentation rate to that in pre-industrial times. Focusing factors (FF in the equation) were estimated for all dated cores by dividing the observed $^{210}Pb$ flux by the predicated $^{210}Pb$ flux for the same latitude, and this predicted $^{210}Pb$ flux in the Xisha Islands is based on the results of Xu et al. (2010). Since we are mainly interested in the change of anthropogenic Hg deposition, we employed the mean Hg flux value before 1850

AD as the background (that is $Hg_B = Hg_{pre\text{-}1850}$, the period from 1450–1550 AD was excluded from the calculation of the $Hg_{pre\text{-}1850}$ due to potential influence from anthropogenic activity), which includes both natural and guano-derived Hg.

## 8.2.2 *Change of Prehistorical Hg Deposition Flux*

The $Hg_A$ fluxes from the three sediment profiles remains pretty low at the time before 1450 AD (Fig. 8.7), and then there is a small peak of anthropogenic Hg flux at approximately 1500 AD. After that, anthropogenic flux remains relatively stable and slightly higher than that before 1450 AD, but significantly lower than its level in the post-industrial era. We suggest that the changes of anthropogenic Hg flux in these three Islands is probably caused by large-scale anthropogenic emission in Asia, especially in recent times after 1850 AD.

Concentrations of Hg in Antarctic penguin faeces and seal hairs were in good agreement with human civilization, and ornithogenic sediments have been identified as ideal materials to reconstruct human metallurgy history (Sun et al. 2006). Although worldwide Hg emission may have potential influences on the concentrations of Hg in the sediments from the Xisha Islands, the impact from the southern hemisphere is probably negligible, due to prevailing direction of atmospheric circulation and that intertropical convergence zone (ITCZ) acts as a natural barrier for its delivery from southern to northern hemisphere. We suggest that the accumulation of Hg in Jinqing, Guangjin and Jinyin Islands is most likely linked with the metallurgical civilizations in the Asia-Pacific region, due to the specific location of the Xisha Islands and impact of the East Asian summer/winter monsoons. Fu et al. (2010) investigated distributions and sources of Hg in the South China Sea, and identified that China and the Indochina peninsula were predominant sources of atmospheric Hg in the northern South China Sea.

The small peaks of adjusted anthropogenic Hg flux at around 1500 AD (Fig. 8.7) correspond to the Yongle Era of the Chinese Ming Dynasty. After the year 1400 AD, Ming China's economy became to recovery, leading to a rapid economic growth and the revival of some heavy industries, e.g. iron and coal mining. During that time, industrial production reached new heights and it exceeded that of the Song Dynasty. For instance, the iron output of iron is approximately three times greater than that of the Song Empire. The booming metallurgical industry made multiple use of silver and gold. The demand for silver was so great that it overwhelmed the earlier paper currency used by the Ming dynasty, and silver soon became the common medium of exchange. Emission from metal smelting was the principal source of Hg, and the utilization of amalgamation was the most popular refining technology, releasing hundreds of thousands of tons of Hg to the atmosphere (Hylander and Meili 2005; Sun et al. 2006). Therefore, the evident Hg peak around 1500 AD is possibly a reflection of the powerful economic activities during the Chinese Ming Dynasty, in parallel with the 2000-year record of copper mining activity from the Xisha Islands (Yan et al. 2010). The economic recession after 1600 AD in Ming Dynasty has been well-known, and it was accompanied by reduced deposition flux of Hg in the sediments (Fig. 8.7). The anthropogenic Hg flux showed a slight rise since 1700 AD, and this might be a result of the massive "gold rush" (1580–1900 AD) in North America and the increased emission of Hg from the Spanish-American silver mines (Camargo 2002). During the "gold rush era", the consumption of Hg exceeded 257,000 tons; especially during 1700–1800 AD, its annual consumption reached as high as 800–1200 tons (Nriagu 1994). Furthermore, with China's political/social stability and economic development, a large quantities of mineral resources were required, and mining activities began to thrive. For example, mining of copper ore reached its peak during the mid-Qing Dynasty (Yan et al. 2010).

### 8.2.3 Recent Change of Hg Deposition Flux

In the early 20th century, more advanced methods like cyanide method (heap leaching) gradually replaced mercury amalgamation; especially wet mercury smelting in the 1940s greatly reduced mercury emission (Lacerda 1997). However, the mercury emissions from coal burning, chlor-alkali industry, oil refining and other activities have become new major sources of atmospheric mercury (Pacyna and Pacyna 2002; Hylander and Meili 2003; Wong et al. 2006; Li et al. 2009). The increasing human mercury consumption has caused the persistent increase of global atmospheric mercury concentration (Slemr and Langer 1992). Before 1850 AD, the total Hg deposition flux on Jinyin, Jinqing and Guangjin islands remained relatively low and stable; the average values were below 40 $\mu g\ m^{-2}\ year^{-1}$, the adjusted anthropogenic Hg flux was no more than 10 $\mu g\ m^{-2}\ year^{-1}$, and anthropogenic Hg was not dominant in total Hg. After 1850 AD, Hg level increased significantly (Fig. 8.7). This is broadly consistent with records from other regions of the world (Sun et al. 2006; Yang et al. 2010a, b; Cooke et al. 2009), reflecting the increasing influences of human activities on the Xisha island ecosystem since the onset of the Industrial Revolution.

We averaged the adjusted anthropogenic mercury deposition fluxes from the three sediment profiles and reconstructed a ~700-year depositional record in order to view the overall historical trend in the Xisha Islands. Because of the large differences in the Hg deposition flux between different time periods, we averaged the adjusted anthropogenic Hg deposition rate at 20-year time intervals from 1400 to 1840 AD, at 10-year time intervals from 1840 to 2000 AD, and at 5-year time interval from 2000 to 2005 AD, based on the method of Muir et al. (2009). The result is presented in Fig. 8.8a. Moreover, we calculated the enrichment factor (EF) or flux ratio (Fig. 8.8a). EF = $F_{recent}/F_{pre\text{-}1850}$, where $F_{recent}$ and $F_{pre\text{-}1850}$ represent Hg flux after 1850 AD and before 1850 AD, respectively. Here, the period from 1450 to 1550 AD was excluded from the calculation of the average Hg flux before 1850 AD due to potential influence from anthropogenic activity. The anthropogenic Hg flux showed a small peak during 1450–1550 AD, corresponding

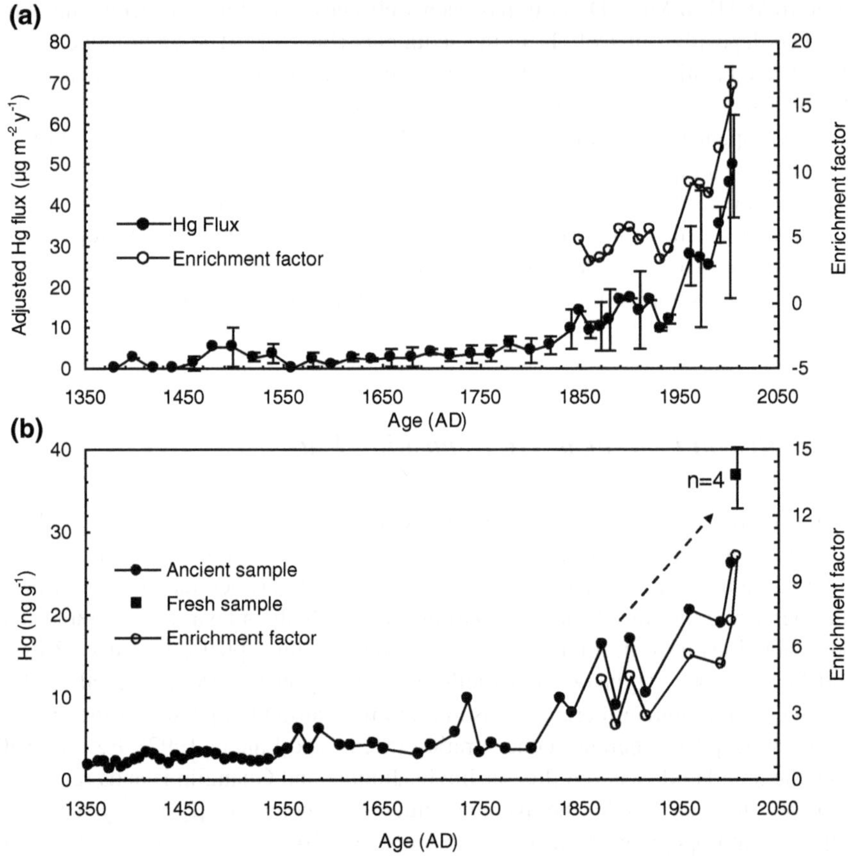


**Fig. 8.8** Average historical change of **a** the adjusted anthropogenic Hg flux for all the three sediment profiles over 5–20 year intervals and **b** the historical changes of Hg concentration in eggshells sampled from GJ3 sediment profile. The *solid line* with *hollow circles* shows the enrichment factor of post-industrial average anthropogenic Hg flux over preindustrial level. The 700-year record of eggshell Hg is after Xu et al. (2011b). [Reprinted with Liu et al. (2012), Copyright (2012), with permission from Elsevier]

to the highly developed metallurgical civilization in ancient China as discussed above. Since 1850 AD, anthropogenic Hg flux had an overall significantly increasing trend, especially at the beginning of the 20th century, with two troughs during the periods around 1940s and 1970s. The first trough corresponds to the World War II, or Anti-Japanese War (1937–1945) in China and Chinese Civil War (1945–1949). This suggests that the Hg flux in the Xisha Islands during that time may be mainly subject to regional anthropogenic activity in China and other Southeast Asia countries, rather than global emissions. The second trough is likely related to Cultural Revolution (1966–1976 AD) in China. Many factories were closed or even destroyed and the economic activities stagnated by this political movement.

The trend of anthropogenic Hg flux in the Yongle archipelago is consistent with those of polycyclic aromatic and other organic pollutants in Chinese coastal sea sediment (such as the Pearl River Delta) (Liu et al. 2005; Guo et al. 2006). It is also consistent with Hg records at different latitudes. Hg fluxes all over the world are relatively low before the Industrial Revolution (Sun et al. 2006; Yang et al. 2010a, b; Cooke et al. 2009). For example, the Hg deposition fluxes of Canada, Sweden, Xisha islands and other regions are all below 10 μg $m^{-2}$ $year^{-1}$ (Muir et al. 2009; Lindeberg et al. 2007). After the Industrial Revolution, the anthropogenic Hg flux increased sharply. According to Muir et al. (2009), the average anthropogenic Hg flux in mid-latitude regions, sub-Arctic, and Arctic regions were 26.8, 7.5, and 2.8 μg $m^{-2}$ $year^{-1}$, respectively. Because of less human influence, anthropogenic Hg flux of high Arctic and subarctic regions are relatively low (Muir et al. 2009). Xisha Islands is located in a low latitude area. The anthropogenic Hg deposition rate of this area began to greatly increase after 1850 AD; current average value is 20.70 μg $m^{-2}$ $year^{-1}$; and it is significantly higher than those of the Arctic and subarctic regions, but still lower than those of the industrialized Americas, Europe and other mid-latitude regions.

The calculated Hg flux ratio (enrichment factor) on Jinyin, Jinqing and Guangjin islands indicated a nearly 7-fold increase in modern Hg fluxes over preindustrial levels. Most studies on the sediment cores from remote lakes showed an average threefold to fivefold increase in global modern atmospheric mercury deposition rates (Lamborg et al. 2002; Biester et al. 2007). Cooke et al. (2009) reported a 4.6-fold increase in Hg accumulation flux from preindustrial to modern times at an equatorial high-altitude site in the Peruvian Andes of South America. Yang et al. (2010b) reported that mercury accumulation in sediment cores from three equatorial zone lakes increased by about threefold since the mid-19th century. The relatively high increase of Hg deposition rate in the Xisha area since preindustrial times might be the result of a sharp increase in atmospheric deposition related to enhanced economic development and human activity in Asia in recent decades.

The industrial usage of Hg has been restricted since late 1970s due to global awareness of Hg toxicity, and worldwide Hg production has been reduced to the 1850 AD level (Hylander and Meili 2003). In European and American industrialized countries, strict measures have been taken, and the Hg deposition flux has been significantly decreased (Lindeberg et al. 2007). However, the modern anthropogenic Hg flux in the Yongle Islands showed a rapid increase and reached the highest value in the recent 40 years. The average enrichment factors of anthropogenic Hg deposition flux is about 12 after 1970 AD, and it peaked at ~16, much higher than 7, the average enrichment factor after 1850 AD. The higher Hg deposition flux of Yongle Islands in recent decades is possibly associated with the increased Hg emissions in the Asia-Pacific countries (Pacyna and Pacyna 2002; Pacyna et al. 2006; UNEP Chemicals Branch 2008; Li et al. 2009; Pirrone et al. 2010), partly due to the migration of Hg production from Europe and America to Asia (especially East Asia and Southeast Asia). Pacyna et al. (2006, 2010) suggested that the Hg emissions of Asia are about 54 and 67 % of the world anthropogenic atmospheric Hg at 2000 AD and 2005 AD, respectively; and of all countries,

China has the largest anthropogenic Hg emission, accounting for 18 %. The anthropogenic Hg emission in China continued to increase at a rate of 2.9 % per year from 1995 to 2003 AD (Wu et al. 2006). Coal burning is the largest anthropogenic source of mercury emissions to the atmosphere (UNEP Chemicals Branch 2008). The total atmospheric emissions of Hg from coal combustion in China have rapidly increased from ~74 tons in 1980 to ~306 tons in 2007 at an annually averaged growth rate of 5.4 % (Tian et al. 2010). Asia has also been the dump site for industrial waste of developed countries (Wong et al. 2006). Recently, the reconstruction of historical Hg records inferred from Chinese lacustrine and marine sediments suggested that atmospheric Hg deposition did increase over the past 30 years (Wu et al. 2008; Xu et al. 2009; Shi et al. 2010; Wang et al. 2010). Therefore, the high Hg production, together with high emission of mercury from coal burning, smelting and waste incineration, of Asia areas in recent decades dominates the remarkable increase in the anthropogenic Hg flux of the Xisha area; this phenomenon is consistent with the Hg flux records in the lake sediments from Tibetan Plateau (Yang et al. 2010a). Streets et al. (2005, 2009) examined Hg emissions in China and predicted global Hg emissions in 2050. They found mercury emissions were high in China and suggested a continued increase in Hg emission in Asia, in consistence with our findings.

High Hg production in Asia can not only contaminate ecological environment, but also cause serious damage to organisms through food chain; therefore, effective monitoring of Hg contamination is very important. Seabirds have a good potential of being used in monitoring Hg and other pollutants in the marine environment. Seabird tissues, including faeces, feathers, and eggshells, have been successfully used as biological indicators of marine Hg pollution (Monteiro and Furness 1995; Kim et al. 1998). We also have analyzed ancient eggshells extracted from ornithogenic sediments on Guangjin Island and reconstructed the historical change of Hg in the eggshells over the past 700 years (Sect. 8.1, Fig. 8.8b). The data showed that the Hg concentration of eggshells increased rapidly after 1800 AD, and the modern eggshell samples had the highest Hg content, averaging 36.7 ng $g^{-1}$ in the four modern samples; this change pattern is very consistent with the one from the Hg flux. Thus, the sediment profiles in the Yongle archipelago provided a good record of the Hg deposition history over the past 700 years. Despite of these similar records, the post-industrial enrichment factors obtained from anthropogenic Hg flux in the present study are greater than those from eggshell Hg record. For example, enrichment factor recorded by modern eggshell samples is <10, while the factor from the bulk sediments in 2008 is approximately 16, indicating that our estimated anthropogenic Hg flux may be overestimated from the direct atmospheric accumulation; and the surface runoff of terrestrial organic matter might also contribute to the recent increase of anthropogenic Hg accumulation at the study sites. The strong relationship between the organic carbon and the sedimentary Hg in the sediments supports this hypothesis. Future study is needed to quantitatively estimate anthropogenic Hg loading from terrestrial organic matter accumulation and direct atmospheric Hg deposition in the sediments of Yongle Islands.

Based on the analysis of Hg content in the ornithogenic sediments from Jinyin, Guangjin and Jinqing Islands, we have reconstructed the historical change of anthropogenic Hg deposition over the past 700 years in the Yongle Islands of South China Sea. Before 1850 AD, anthropogenic Hg flux was relatively low with a small peak around 1450–1550 AD, possibly due to enhanced metallurgical activities during China Ming Dynasty. Anthropogenic Hg flux increased rapidly after the Industrial Revolution, but with two periods of low values. The first period of 1940s is likely linked with the World War II and China Civil war. The second period of 1970s is likely associated with the Culture Revolution of China. The calculated Hg flux ratio (enrichment factor) indicated a nearly sevenfold increase since preindustrial times. After 1970s, the mercury flux in Xisha Islands increased sharply, and the present enrichment factors is ~16. This is likely attributed to the increases of both terrestrial organic matter accumulation and anthropogenic Hg emissions in the Asia-Pacific countries, partly due to the migration of Hg production from Europe and America to Asia. As the Hg pollution is expected to increase continuously, more attention toward Hg and other heavy metal pollution is needed in order to protect the integrity of coral island ecosystem.

## 8.3 A 400-Year Record of Black Carbon Flux in the Xisha Archipelago, South China Sea and Its Implication

### *8.3.1 Introduction to Black Carbon (BC)*

Black carbon (BC) is a class of carbon compounds produced mainly by incomplete combustion of biofuel, fossil fuels or biomass (Masiello 2004). Such compounds are chemically heterogeneous and are widespread in various natural archives and materials, for example air, soil, sediment, water/snow. BC is chemically stable and thus can be well-preserved even after hundreds of years (Goldberg 1985; Verardo and Ruddiman 1996; Kuhlbusch 1998; Masiello and Druffel 1998; Schmidt and Noack 2000; Gelinas et al. 2001; Jacobsen 2002; Muir et al. 2002, 2006; Fernandes et al. 2003; Shrestha et al. 2010). BC in sediments has a strong resistance to microbes and is a robust marker for changes in fire utilization. It has been frequently used to reconstruct the history of forest exploitation and widefire, as well as the gradual changes of energy structure (Verardo and Ruddiman 1996; Marlon et al. 2008; Sun et al. 2008). Moreover, BC acts as a sink for persistent organic matter by adsorption, and plays a significant role in biogeochemical cycling of some elements, especially for oxygen and carbon (Kuhlbusch and Crutzen 1995; Menon et al. 2002; Bucheli et al. 2004; Sánchez-García et al. 2010). Consequently, the study of BC in environmental archives has bee a hot research topic in recent years.

With the expansion of human civilization, i.e. the Industrial Revolution, a great deal of BC has been emitted into the atmosphere. This led to a marked increase of its background level. Rapid economic development in Asia in recent times, especially over the past several decades, has rendered this region the foremost source of anthropogenic contaminants (Streets et al. 2003; Pacyna et al. 2010). Anthropogenic BC was released into the atmosphere, transported to remote areas via a long-distance atmospheric circulation, brought to the Earth surface by wet or dry precipitations, and eventually entered soils or sediments. Many studies have reconstructed deposition history of BC in marine sediments, and it was found that both river transportation and human activities imposed significant influences on distributions of BC in such oceanic sediments (Middelburg et al. 1999; Jia et al. 2000; Wang and Li 2007; Sun et al. 2008; Ribeiro et al. 2008; Kang et al. 2009; González-Vila et al. 2009). However, most of previous studies used marine sediments as study materials; and none of previous research focused on distribution and biogeochemical cycling of BC on coral islands, which do not receive any terrestrial fluvial input.

In this part, we attempt to determine the levels of BC in our guano sediments from Jinqing, Guangjin and Jinyin Islands. It is aimed to identify the dominant source and distribution features of BC in the ornithogenic sediments, to reconstruct deposition flux of BC on the Xisha archipelago over the past four centuries, and to examine its potential environmental significance.

### *8.3.2 Analytical Methods*

Following the method by Lim and Cachier (1996), BC in bulk sediments was extracted as follows. An aliquot of sample was precisely taken and place to a beaker. Sediment carbonate was removed by excess 1:1 HCl, and then HF and HCl were added for approximately 12 h until all the silicate was completely removed. To further eliminate calcium ion and fluoride, the solid residues were washed with distilled water for several times. 10 mL $K_2Cr_2O_7$ (0.1 mol/L) and $H_2SO_4$ (2 mol/L) was then added to react for 60 h. After centrifuging and drying at a constant temperature of 105 °C for 12 h, BC concentration was determined by elemental analyzer (Vario EL III). We also performed SEM (XL30 ESEM-TMP, Phillips) analysis to obtain micrographs of black carbon particles.

### *8.3.3 Vertical Distributions of Black Carbon in the Three Sediment Profiles*

Concentration-versus-depth profiles of BC in profiles GJ3, JQ and JY2 were supplied in Fig. 8.9. The ranges of BC in three ornithogenic sediment cores were 0.054–1.04 mg $g^{-1}$ (GJ3), 0.34–1.44 mg $g^{-1}$ (JY2) and 0.016–2.95 mg $g^{-1}$ (JQ), respectively, and the average BC values in the cores were 0.31 mg $g^{-1}$ ($n = 29$, GJ3), 0.73 mg $g^{-1}$ ($n = 12$, JY2) and 0.42 mg $g^{-1}$ ($n = 14$, JQ). Though TOC and BC fluctuated significantly throughout the three profiles, their concentrations

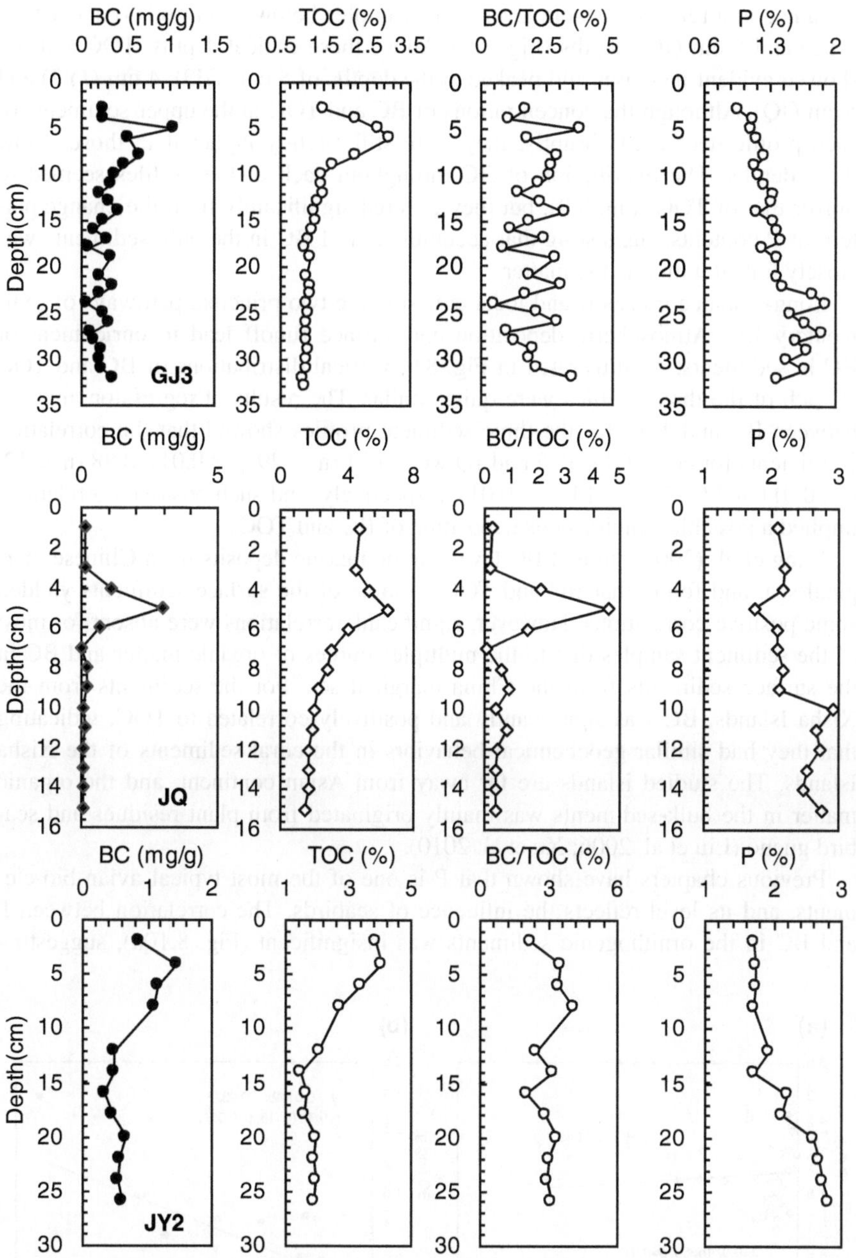


**Fig. 8.9** Contents of BC, TOC, BC/TOC and P versus depth in GJ3, JQ and JY2 profiles [Reprinted from Liu et al. (2011b), Copyright (2011), with permission from Elsevier]

remained at a relatively stable level in the samples below 12 cm (GJ3) and 10 cm (JY2) and 8 cm (JQ) depths (Fig. 8.9). Above these critical depths, TOC and BC showed evident increases and peaked at the depths of 8 cm (GJ3), 4 cm (JY2) and 5 cm (JQ). Although the concentrations of BC and TOC in the upper sediments of each profile decreased abruptly, they were still slightly higher than those below these depths. The distribution of BC throughout each of the profiles seemed to mirror that of TOC (Fig. 8.9); but they differed significantly from the change pattern of P contents, suggesting that accumulation of BC in the bulk sediments was closely linked with organic matter.

Long-distance transport and local emission are two principal pathways of sedimentary BC. Atmospheric deposition and surface runoff lead to enrichment of BC in sediments. As illustrated in Fig. 8.9, vertical distributions of BC and TOC in each of the three profiles were quite similar. The results of regression analysis between BC and TOC for the three sediment profiles showed that the correlation coefficients for cores GJ3, JY2 and JQ were 0.74 ($n = 29, p < 0.01$), 0.88 ($n = 12, p < 0.01$) and 0.75 ($n = 14, p < 0.01$), respectively, and such positive correlations implied a possible simultaneous deposition of BC and TOC.

Kang et al. (2009) studied BC levels in the marine deposits from Chinese marginal sea, and found that BC and TOC in some of the surface sediments yielded some positive correlations. However, significant correlations were absent for most of the sediment samples due to the multiple sources of organic matter and BC in the surface sediments from the China marginal sea. For the sediments from the Xisha Islands, BC was significantly and positively correlated to TOC, indicating that they had similar geochemical behaviors in the coral sediments of the Xisha Islands. The studied islands are far away from Asian continent, and the organic matter in the bulk sediments was mainly originated from plant residues and seabird guano (Liu et al. 2006; Xu et al. 2010).

Previous chapters have shown that P is one of the most typical avian bio-elements, and its level reflects the influence of seabirds. The correlation between P and BC in the ornithogenic sediments was insignificant (Fig. 8.10a), suggesting

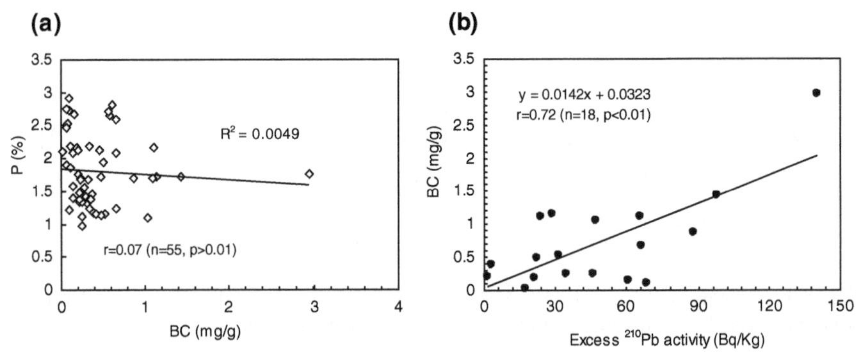


**Fig. 8.10** Correlation between BC, P and excess $^{210}$Pb in GJ3, JQ and JY2 [Reprinted from Liu et al. (2011b), Copyright (2011), with permission from Elsevier]

that guano is not a major source for sedimentary BC. According to analysis of radionuclides in the sediments, it was found that excess Pb-210 ($^{210}Pb_{ex}$) in the ornithogenic sediments was from atmospheric deposition, rather than guano input or local emission (see Chap. 4 and Xu et al. 2010). If BC in the sediments was from atmospheric transport, we would expect a significant correlation between excess $^{210}Pb$ and BC concentration. Our speculation was confirmed by Pearson correlation analysis (Fig. 8.10b), suggesting that both BC and $^{210}Pb_{ex}$ were sourced from atmospheric dry/wet deposition and further retained by sediment organics. Therefore, we suggest that BC in the ornithogenic coral sand sediments, like radionuclides (Xu et al. 2010), is mainly from atmospheric precipitation; the organic matter could adsorb and thereby enhance the BC levels in the sediments or soils due to its strong affinity with organic matter.

The amount of BC from a specific source may vary significant, but the ratio of BC/TOC generally remains stable and thus has been widely used in source identification (Muir et al. 2002; Sun et al. 2008; He and Zhang 2009). The values of BC/TOC in GJ3, JQ and JY2 were less than 5 % (Fig. 8.9), similar to the values of 4–8 % in remote area sediments (Muir et al. 2002), and were significantly lower than those in both Chinese margin sea sediments and urban soils. Kang et al. (2009) reported that the range of BC/TOC in the surface sediments of northern Yellow Sea, Jiaozhou Bay, East and South China Sea, was 5–14 %, and even up to 27–41 % in sediments from Bohai Bay. Wang and Li (2007) analyzed BC levels in the sediments from continental shelf of East China Sea and Yangtze River Delta, and the results showed that the ranges of BC/TOC in these areas were 5–15 % and 17–26 %, respectively. Jiang et al. (2010) found the BC/TOC was 16.8 ± 7.5 % in the surface sediment from coastal zone in Bohai Bay. Moreover, BC/OC in soil profiles from urban areas in Nanjing of China has a mean of 29 % (He and Zhang 2009). Additionally, several studies have reported relatively high values of BC/TOC in the harbor and gulf sediments from other countries, i.e. 6.4–12.3 % in Norway (Oen et al. 2006), 16–18 % in Boston of USA (Accardi-Dey and Gschwend 2002), 2.3–12.8 % in Guanabara of Brazil (Ribeiro et al. 2008), and 3.0–16.5 % in Cadiz of Spain (González-Vila et al. 2009). Compared with marginal sea sediments and urban soils, the BC/TOC in the sediments of Xisha archipelago was markedly low. This suggested that anthropogenic BC had less influence on the Xisha archipelago than on marginal sea areas.

### *8.3.4 BC Fluxes in the Xisha Archipelago Over the Past 400 Years*

Generally, BC is mainly derived from industrial activity, traffic pollution, outdoor incineration, household coal consumption and biomass. In some of the developing countries, lumber, field residual, cow dung and coal are still used as the principal energy source for cooking and heating. Incomplete combustion of these materials at low temperature would generate large amount of BC. It was estimated that China released up to 41 % of BC in Asia, and incomplete combustion of unprocessed briquette, coal and biomass were the foremost contribution to BC emission in China (Streets et al. 2003; Cao et al. 2006; Liu and Shao 2007; Ohara et al. 2007). Streets et al. (2001, 2003) reported that BC released by China in 1995 and 2000 were 1.34 Tg and 1.05 Tg, respectively, 60 % of which were from incomplete combustion of household coal and biomass. BC/TOC was a robust and widely accepted marker for BC source (He and Zhang 2009). Incomplete combustion of biomass will generate a BC/TOC value below 0.11, while BC/TOC produced by fossil fuels is approximately 0.5. Based on Fig. 8.9, we observed that BC/TOC in the three sediment profiles were all far below 0.11 and this suggested that BC in the sediments of Xisha Islands was principally attributed to the incomplete combustion of biomass.

To eliminate the influence of sedimentation rate and soil organic matter on BC distribution, we followed the method of Sun et al. (2008) and calculated the BC fluxes of GJ3, JQ1 and JY2 profiles using the concentration data, the dry bulk density and sediment accumulation rate estimated from the respective $^{210}$Pb-AMS$^{14}$C age models. The reconstructed BC fluxes in the Xisha Islands over the past approximately 400 years were shown in Fig. 8.11a. The ranges and temporal distributions of BC flux of the three islands were quite similar. In order to view the whole change trend of the historical BC deposition flux in the Xisha archipelago, we processed the data at different time stages. The BC deposition rate was averaged over 20-year time intervals from 1650 to 1950 AD, 10-year time intervals from 1950–2000 AD, and the result is presented in Fig. 8.11b. As shown in Fig. 8.11, BC flux stayed at low level before 20th century and started to increase rapidly from then on. After reaching the peaks at around 1970s, BC fluxes seemed to display decrease trend in the Xisha Islands. These results were consistent with the BC flux record in the sediments of Pearl River Delta, South China Sea (Sun et al. 2008).

The mean value of BC flux in the three profiles was 0.027 mg cm$^{-2}$ year$^{-1}$ before 1850. In the 1970s, BC fluxes reached their peak values of 0.09 (GJ3), 0.1–0.14 (JQ) and 0.09–0.11 (JY2) mg cm$^{-2}$ year$^{-1}$. Like BC/TOC, the BC flux in the Xisha Islands was similar to that in remote areas, but much lower than that in inshore and urban regions. For example, the BC flux in the remote Alps started at 0.03 mg cm$^{-2}$ year$^{-1}$ before 1850, increased to 0.05–0.62 mg cm$^{-2}$ year$^{-1}$ in modern times, and decreased further in recent decades (Muir et al. 2002). By contrast, Sun et al. (2008) reported that the ranges of BC fluxes were 0.69–2.56 and 0.55–1.73 mg cm$^{-2}$ year$^{-1}$ in two sediment cores retrieved from estuary of Pearl River, much higher than our results despite the geographical proximity. Gustafsson and Gschwend (1998) studied the BC distribution in surface sediments in the marginal area of NE America continental shelf; the BC flux in harbor area nearby cities was the highest up to 0.5–2 mg cm$^{-2}$ year$^{-1}$, and the BC flux in Maine Gulf (nonurban Gulf of Maine shelf stations) far from the urban area was only 0.1–0.2 mg cm$^{-2}$ year$^{-1}$, close to that in the Xisha Islands. Anthropogenic BC originated from fuel usage could be transported worldwide due to global atmospheric

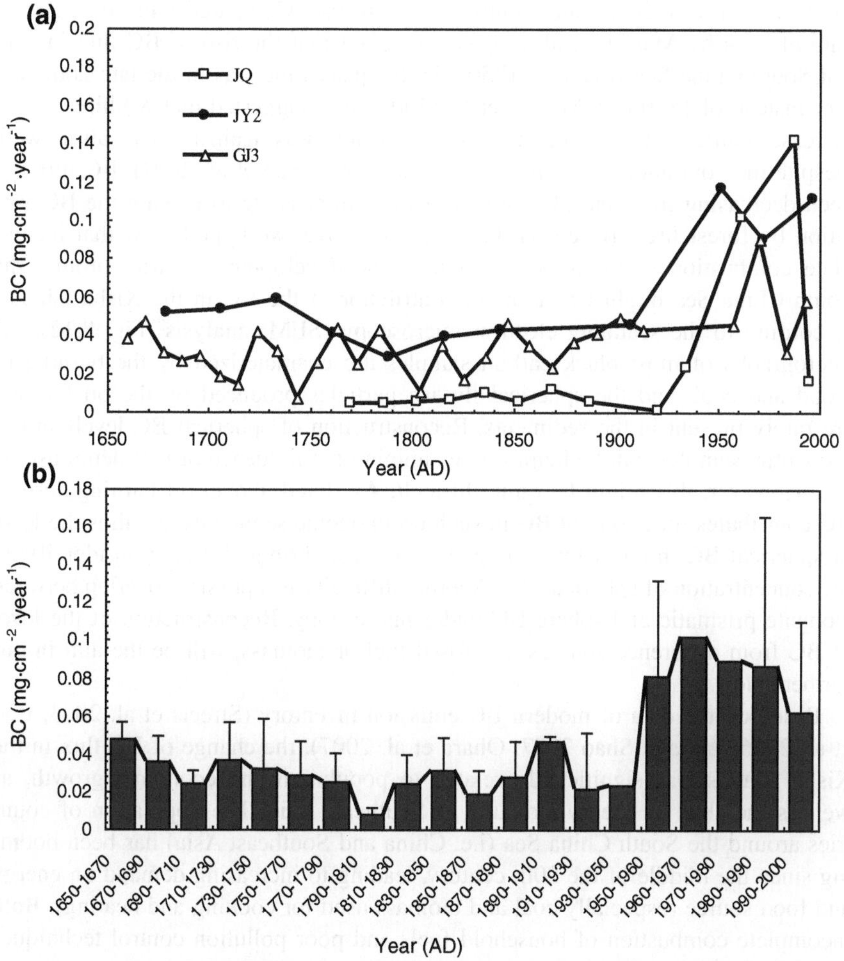


**Fig. 8.11** Record of BC deposition flux in GJ3, JQ and JY2 profiles over the past 400 years [Reprinted from Liu et al. (2011b), Copyright (2011), with permission from Elsevier]

circulation; but the flux in different places of the world may be quite different, and the deposited BC flux in remote and isolated areas, e.g. the Xisha Islands, tends to be low.

The changes of BC flux in the Xisha Island cannot be explained by forest fires. Marlon et al. (2008) collected charcoal samples worldwide and reconstructed the records of both natural and anthropogenic biomass combustion over the past 2,000 years. They found that global biomass combustion gradually decreased from 1 to 1750 AD, steadily increase during 1750 and 1870 AD, and sharply dropped afterward. The first period of decrease corresponded to the global cooling, and the following increase was related to population growth and land use. The decrease of biomass combustion in the recent 150 years is attributed to overgrazing, global expansion of agriculture, and control upon fire use. Compared with the charcoal data of Asia by Marlon et al. (2008), we found that the rise of BC flux in central South China Sea was after 1750 AD; the peak time was in the late 20th century instead of 1870 AD. Marlon et al. (2008) also suggested that wildfires would increase with continuous climate warming, which was against our result as well. Despite the continuous warming in the Xisha area (Sun et al. 2004), BC flux has been decreasing in recent 30 years, making it impossible to explain the BC variation by forest fires. Based on the evidence above, we hypothesize that incomplete combustion of biomass and coal in the developing countries around the South China Sea might be the main contributor of the BC in the Xisha Islands. According to the scanning electron microscopy (SEM) analysis (Fig. 8.12), the micrographs of most black carbon samples are characterized by the burning of wood and coal, and the spherical fly-ash particles produced by the oil burning are rarely present in the sediments. Reconstruction of spherical BC levels in our sequential samples will be helpful in examining the influence of petroleum use on BC. However, this might be quite difficult. As stated above, oil burning-derived BC contributes little to total BC in such ornithogenic sediments and thus the level of spherical BC in our samples is pretty low. This brings difficulty in identifying the concentration of spherical BC. Another difficulty is a possible overlap between elongate prismatic and sphere BC under microscopy. Reconstruction of the level of BC from difference sources, i.e. fossil fuel or biomass, will be the aim in our further study.

Based on the data of modern BC emission inventory (Streets et al. 2003; Cao et al. 2006; Liu and Shao 2007; Ohara et al. 2007), the change of BC flux in the Xisha Islands was significantly related to population and economic growth, as well as the shift of energy structure in Southeast Asia. The population of countries around the South China Sea (i.e. China and Southeast Asia) has been booming since the middle of the 20th century, leading to increasing demand on energy and food source (especially coal and biomass used for cooking and heating). Both incomplete combustion of household fuels and poor pollution control techniques

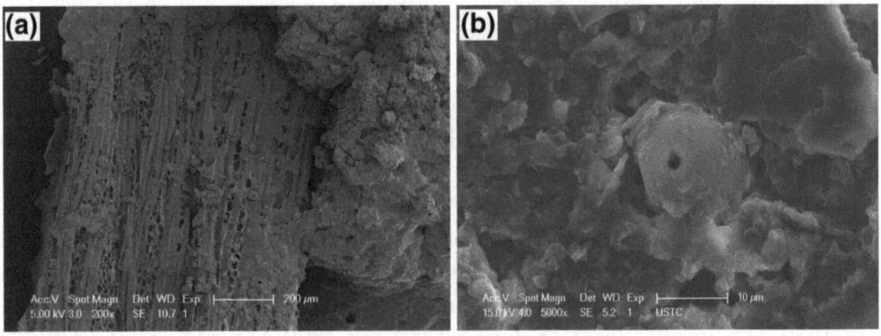


**Fig. 8.12** Scanning electron micrographs of black carbon samples from the burning of biomass (**a**) and coal (**b**) [Reprinted from Liu et al. (2011b), Copyright (2011), with permission from Elsevier]

resulted in increasing release of BC, which was further transported by the atmosphere and deposited on the Xisha Islands.

The BC deposition in the Xisha Islands peaked in 1970s. Since 1980s, The BC deposition decreased gradually as rapid industrialization and urbanization occurred in southern China (Pearl River Delta for example). Two factors may have contributed to the decreased BC deposition. First is the improvement in combustion techniques and pollution, together with energy structure shift from mono-dominant biomass and coal fuel to a mixed structure of coal, natural gas, petroleum and biomass. Sun et al. (2008) reported a similar change of historical BC flux record in the Pearl River estuary and adjacent northern South China Sea. Second is the fast urbanization as a large quantity of farmland was transformed into cities (Roberts and Chan 1997). With the rapid urbanization in countries around the South China Sea since 1980s, the demand for fossil fuels such as petroleum and coal increased significantly. Wu et al. (2005) reported that the level of spheroidal carbonaceous particle (SCP) in the lacustrine sediments of central China, which was formed from combustion of fossil fuel at high temperature up to 1,750 °C, had increased significantly since 1978, suggesting an increased use of fossil fuel. Generally, fossil fuels used in industrial production release less BC than the usage of biomass fuel for small household stove (Sun et al. 2008). At the same time, consumers also prefer fossil fuels and clean energy to biomass. Streets et al. (2003) reported that BC released by China could drop by 8 % from 1995–2000 AD, contrasting with the significant increase of total demand for fossil energy, and the reduction was attributed to the use of more advanced fuel technologies. For example, the processed honeycomb briquette and natural gas were used to substitute unprocessed coal and biomass fuel in the household stove; the proportion of alternative electricity and other clean energy increased significantly. We have to point out here that the present level of BC deposition is still significantly higher than that in the 1900s (Fig. 8.11).

### *8.3.5 Brief Summary*

Time series of black carbon levels in the ornithogenic sediments have been examined. The significant correlation between BC and excess $^{210}$Pb and the lack of correlation between BC and P in the three sediment cores of Xisha Islands suggested that BC in the ornithogenic coral sand sediments originated mainly from atmospheric deposition, rather than guano input. BC was brought down to the earth surface by dry or wet precipitation, and then the organic matter enhanced the ability of sediments to retain it. The anthropogenic BC in Xisha Islands was principally derived from incomplete combustion of biomass and coal by household cooking and heating in developing countries of East and Southeast Asian, but its impact on the sediments of the Xisha Islands was insignificant. The BC flux in the Xisha Islands was low before the onset of 20th century, it had a rapid increase since the middle of twentieth century, and peaked in 1970s. The gradual decrease of BC in the past 30 years possibly reflected the change of energy structure from biomass to fossil fuels and the improvement of pollution control technique in the regions around the South China Sea.

## References


Accardi-Dey A, Gschwend PM (2002) Assessing the combined roles of natural organic matter and black carbon as sorbents in sediments. Environ Sci Technol 36:21–29

Biester H, Bindler R, Martinez-Cortizas A, Engstrom DR (2007) Modeling the past atmospheric deposition of mercury using natural archives. Environ Sci Technol 41:4851–4860

Bucheli TD, Blum F, Desaules A, Gustafsson Ö (2004) Polycyclic aromatic hydrocarbons, black carbon, and molecular marker in soils of Switzerland. Chemosphere 56:1061–1076

Camargo JA (2002) Contribution of Spanish-American silver mines (1570–1820) to the present high mercury concentrations in the global environment: a review. Chemosphere 48:51–57

Cao L (2005) Population ecology of the red-footed booby on the Xisha Archipelago. PhD dissertation. Lanzhou University

Cao GL, Zhang XY, Wang YQ, Che HZ, Chen D (2006) Inventory of black carbon emission from China. Adv Clim Change Res 2(6):259–264 (in Chinese with English abstract)

Cooke CA, Balcom PH, Biester H, Wolfe AP (2009) Over three millennia of mercury pollution in the Peruvian Andes. Proc Natl Acad Sci USA 106:8830–8834

Dai SG (1997) Environmental chemistry. Higher Education Press, Beijing

Dietz R, Outridge PM, Hobson KA (2009) Anthropogenic contributions to mercury levels in present-day Arctic animals—a review. Sci Total Environ 407:6120–6131

Emslie SD, Patterson WP (2007) Abrupt recent shift in $\delta^{13}C$ and $\delta^{15}N$ values in Adelie penguin eggshell in Antarctica. Proc Natl Acad Sci 104(28):11666–11669

Fernandes MB, Skjemstad JO, Johnson BB, Wells JD, Brooks P (2003) Characterization of carbonaceous combustion residues. I. Morphological, elemental and spectroscopic features. Chemosphere 51:785–795

Feyte S, Tessier A, Gobeil C, Cossa D (2010) In situ adsorption of mercury, methylmercury and other elements by iron oxyhydroxides and organic matter in lake sediments. Appl Geochem 25:984–995

Fitzgerald WF, Engstrom DR, Mason RP, Nater EA (1998) The case for atmospheric mercury contamination in remote areas. Environ Sci Technol 32:1–7

Fu X, Feng X, Zhang G, Xu W, Li X, Yao H, Liang P, Li J, Sommar J, Yin R, Liu N (2010) Mercury in the marine boundary layer and seawater of the South China Sea: concentrations, sea/air flux, and implication for land outflow. J Geophys Res 115:D06303. doi:10.1029/2009JD012958

Gelinas Y, Prentice KM, Baldock JA, Hedges JI (2001) Improved thermal oxidation method for the quantification of soot/graphitic black carbon in sediments and soils. Environ Sci Technol 35:3519–3525

Gil C, Ramos-Miras J, Roca-Pérez L, Boluda R (2010) Determination and assessment of mercury content in calcareous soils. Chemosphere 78:409–415

Goldberg ED (1985) Black carbon in the environment. Wiley, New York, pp 1–146

González-Vila FJ, de la Rosa JM, González-Pérez JA (2009) Black carbon and other refractory forms in recent sediments from the Gulf of Cadiz, Spain. In: IOP conference series: earth and environmental science, vol 5, p 012009

Guo Z, Lin T, Zhang G, Yang Z, Fang M (2006) High-resolution depositional records of polycyclic aromatic hydrocarbons in the central continental shelf mud of the East China Sea. Environ Sci Technol 40:5304–5311

Gustafsson O, Gschwend PM (1998) The flux of black carbon to surface sediments on the New England continental shelf. Geochim Cosmochim Ac 62:465–472

He Y, Zhang GL (2009) Historical record of black carbon in urban soils and its environmental implications. Environ Pollut 157:2684–2688

Hong SM, Lee K, Hou S, Hur SD, Ren J, Burn LJ, Rosman KJ, Barbante C, Bountron CF (2009) An 800-year record of atmospheric As, Mo, Sn, and Sb in Central Asia in high-altitude ice cores from Mt. Qomolangma (Everest), Himalayas. Environ Sci Technol 43(21):8060–8065

Hylander LD, Meili M (2003) 500 years of mercury production: global annual inventory by region until 2000 and associated emissions. Sci Total Environ 304:13–27

Hylander LD, Meili M (2005) The rise and fall of mercury: Converting a resource to refuse after 500 years of mining and pollution. Environ Sci Technol 35:1–36

Jacobsen MZ (2002) Control of fossil-fuel particulate black carbon and organic matter, possibly the most effective method of slowing global warming. J Geophys Res 107:4410–4431

Jia G, Peng P, Sheng G, Fu J (2000) Sedimentary records of black carbon in the sea area of the Nansha Islands since the last glaciations. Chin Sci Bull 45(17):1594–1598

Jiang GB, Shi JB, Feng XB (2006) Mercury pollution in China. Environ Sci Technol 40(12):3672–3678

Jiang XH, Chen YJ, Tang JH, Huang GP, Liu DY, Li J, Zhang G (2010) The distribution of black carbon in the surface sediments of coastal zone, Bohai Bay. Ecol Environ Sci 19(7):1617–1621 (in Chinese with English abstract)

Kamman NC, Engstrom DR (2002) Historical and present fluxes of mercury to Vermont and New Hampshire lakes inferred from $^{210}$Pb dated sediment cores. Atmos Environ 36:1599–1609

Kang Y, Wang X, Dai M, Feng H, Li A, Song Q (2009) Black carbon and polycyclic aromatic hydrocarbons (PAHs) in surface sediments of China's marginal seas. Chin J Oceanol Limnol 27(2):297–308

Kennamer RA, Stout JR, Colwell SV, Brisbin ILJR, Burger J (2005) Mercury patterns in wood duck eggs from a contaminated reservoir in South Carolina, USA. Environ Toxicol Chem 24(7):1793–1800

Kim EY, Goto R, Tanabe S, Tanaka H, Tatsukawa R (1998) Distribution of 14 elements in tissues and organs of oceanic seabirds. Arch Environ Con Tox 35:638–645

Kuhlbusch TAJ (1998) Black carbon and the carbon cycle. Science 280:1903–1904

Kuhlbusch TAJ, Crutzen PJ (1995) Toward a global estimate of block carbon in residues of vegetation fires representing a sink of atmospheric $CO_2$ and a source of $O_2$. Global Biogeochem Cycle 9:491–501

Lacerda LD (1997) Global mercury emissions from gold and silver mining. Water Air Soil Pollut 97:209–221

Lam JCW, Tanabe S, Wong BSF, Lam PKS (2004) Trace element residues in eggs of Little Egret (Egretta garzetta) and Black-crowned Night Heron (Nycticorax nycticorax) from Hong Kong, China. Mar Pollut Bull 48:378–402

Lamborg CH, Fitzgerald WF, Damman AWH, Benoit JM, Balcom PH, Engstrom DR (2002) Modern and historic atmospheric mercury fluxes in both hemispheres: global and regional mercury cycling implications. Global Biogeochem Cycle 16:1104. doi:10.1029/2001GB1847

Li P, Feng XB, Qiu GL, Shang LH, Li ZG (2009) Mercury pollution in Asia: a review of the contaminated sites. J Hazard Mater 168:591–601

Lim B, Cachier H (1996) Determination of black carbon by chemical oxidation and thermal treatment in recent marine and lake sediments and cretaceous-tertiary clays. Chem. Geol. 131:143–154

Lindeberg C, Bindler R, Bigler C, Rosén P, Renberg I (2007) Mercury pollution trends in subarctic lakes in the northern Swedish mountains. Ambio 36:401–405

Liu Y, Shao M (2007) Estimation and prediction of black carbon emissions in Beijing City. Chin Sci Bull 52:1274–1281

Liu GQ, Zhang G, Li XD, Li J, Peng XZ, Qi SH (2005) Sedimentary record of polycyclic aromatic hydrocarbons in a sediment core from the Pearl River Estuary, South China. Mar Pollut Bull 51:912–921

Liu XD, Zhao SP, Sun LG, Yin XB, Wang YH, Liu KX, Wu XH (2006) Geochemical evidence for the variation of historical seabird population on the Dongdao Island of South China Sea. J Paleolimnol 36:259–279
Liu XD, Sun LG, Yin XB, Wang YH (2008) Heavy metal distributions and source tracing in the lacustrine sediments of Dongdao Island, South China Sea. Acta Geologica Sinica 82(5):1002–1014
Liu XD, Sun J, Sun LG, Liu WQ, Wang YH (2011a) Reflectance spectroscopy: a new approach for reconstructing penguin population size from Antarctic ornithogenic sediments. J Paleolimnol 45:213–222
Liu XD, Xu LQ, Sun LG, Liu F, Wang YH, Yan H, Liu Y, Luo YH, Huang J (2011b) Marine Pollut Bull 62:2205–2212
Liu XD, Xu LQ, Chen QQ, Sun LG, Wang YH, Yan H, Liu Y, Luo YH, Huang J (2012) Historical change of mercury pollution in remote Yongle archipelago. Chemosphere 87:549–556
Lockhart WL, Macdonald RW, Outridge PM, Wilikinson P, Delaronde JB, Rudd JWM (2000) Tests of the fidelity of lake sediment core records of mercury deposition to known histories of mercury contamination. Sci Total Environ 260:171–180
Ma CX, Zhou XP, Chen XQ, Li Q, Lin QX (2005) The study of ultra-structure and element composition of eggshells in *Egretta eulophotes*. J Xiamen Univ (Nat Sci) 44(6):861–865
Marlon JR, Bartlein PJ, Carcaillet C, Gavin DG, Harrison SP, Higuera PE, Joos F, Power MJ, Prentice IC (2008) Climate and human influences on global biomass burning over the past two millennia. Nat Geosci 1:697–702
Martínez-Cortizas A, Pontevedra-Pombal X, García-Rodeja E, Novoa-Muñoz JC, Shotyk W (1999) Mercury in a Spanish Peat Bog: archive of climate change and atmospheric metal deposition. Science 284:939–942
Masiello CA (2004) New direction in black carbon organic geochemistry. Mar Chem 92:201–213
Masiello CA, Druffel EM (1998) Black carbon in deep-sea sediment. Science 280:1911–1913
Menon S, Hansen J, Nazarenko L, Luo Y (2002) Climate effects of black carbon aerosols in China and India. Science 297:2250–2253
Middelburg JJ, Nieuwenhuize J, van Breugel P (1999) Black carbon in marine sediments. Mar Chem 65:245–252
Monteiro LR, Furness RW (1995) Seabirds as monitors of mercury in the marine environment. Water Air Soil Pollut 80:851–870
Muir DCG, Wang X, Yang F, Nguyen N, Jackson TA, Evans MS, Douglas MSV, Köck G, Lamoureux S, Pienitz R, Smol JP, Vincent WF, Dastoor A (2009) Spatial trends and historical deposition of mercury in eastern and northern Canada inferred from lake sediment cores. Environ Sci Technol 43:4802–4809
Muir G, Cermelj B, Faganeli J, Brancelj A (2002) Black carbon in Slovenian alpine lacustrine sediments. Chemosphere 46:1225–1234
Muri G, Wakeham SG, Rose NL (2006) Records of atmospheric delivery of pyrolysis-derived pollutants in recent mountain lake sediments of the Julian Alps (NW Slovenia). Environ Pollut 139:461–468
Nriagu JO (1994) Mercury pollution from the past mining of gold and silver in the America. Sci Total Environ 149:167–181
Nys Y, Gautron J (2007) Structure and formation of eggshell. In: Rainer H, Rosina L-F, Marc A, Rüdiger S (eds) Bioactive egg compounds. Springer, Berlin
Oen AMP, Cornelissen G, Breedveld GD (2006) Relation between PAH and black carbon contents in size fractions of Norwegian harbor sediments. Environ Pollut 141:370–380
Ohara T, Akimoto H, Kurokawa J, Horii N, Yamaji K, Yan X, Hayasaka T (2007) An Asian emission inventory of anthropogenic emission sources for the period 1980–2020. Atmos Chem Phys Discuss 7:6843–6902
Outridge PM, Sanei H, Stern GA, Hamilton PB, Goodarzi F (2007) Evidence for control of mercury accumulation rates in Canadian high arctic lake sediments by variations of aquatic primary production. Environ Sci Technol 41:5259–5265

Pacyna EG, Pacyna JM (2002) Global emission of mercury from anthropogenic sources in 1995. Water Air Soil Pollut 137:149–165

Pacyna EG, Pacyna JM, Steenhuisen F, Wilson S (2006) Global anthropogenic mercury emission inventory for 2000. Atmos Environ 40:4048–4063

Pacyna EG, Pacyna JM, Sundseth K, Munthe J, Kindbom K, Wilson S, Steenhuisen F, Maxson P (2010) Global emission of mercury to the atmosphere from anthropogenic sources in 2005 and projections to 2020. Atmos Environ 44:2487–2499

Perry E, Norton SA, Kamman NC, Lorey PM, Driscoll CT (2005) Deconstruction of historic mercury accumulation in lake sediments, Northeastern United States. Ecotoxicology 14:85–99

Phillips VJ, Louis VS, Cooke CA, Vinebrooke RD, Hobbs WO (2011) Increased mercury loading to western Canadian alpine lakes over the past 150 years. Environ Sci Technol 45:2042–2047

Pietrelli L, Biondi M (2009) Notes on Little Egret breeding biology and on mercury content in egg shells and feather. Rend. Fis. Acc. Lincei 20:219–224

Pirrone N, Cinnirella S, Feng X, Finkelman RB, Friedli HR, Leaner J, Mason R, Mukherjee AB, Stracher GB, Streets DG, Telmer K (2010) Global mercury emissions to the atmosphere from anthropogenic and natural sources. Atmos Chem Phys Discuss 10:4719–4752

Post DM (2002) Using stable isotopes to estimate trophic position: models, methods and assumptions. Ecology 83:703–718

Ravichandran M (2004) Interactions between mercury and dissolved organic matter—a review. Chemosphere 55:319–331

Renberg I (1986) Concentration and annual accumulation values of heavy metals in lake sediments: their significance in studies of the history of heavy metal pollution. Hydrobiologia 143:379–385

Ribeiro LGL, Carreira RS, Wagener ALR (2008) Black carbon contents and distribution in sediments from the southeastern Brazilian coast (Guanabara Bay). J Braz Chem Soc 19:1277–1283

Roberts P, Chan R (1997) A tale of two regions: strategic planning for sustainable development in east and west. Int Plann Stud 2:45–62

Roos-Barraclough F, Martinez-Cortizas A, García-Rodeja E, Shotyk W (2002) A 145,000 year record of the accumulation of atmospheric mercury in peat: volcanic signals, anthropogenic influences and a correlation to bromine accumulation. Earth Planet Sci Lett 202:435–451

Sánchez-García L, Cato I, Gustafsson Ö (2010) Evaluation of the influence of black carbon on the distribution of PAHs in sediments from along the entire Swedish continental shelf. Mar Chem 119:44–51

Sanei H, Outridge PM, Dallimore A, Hamilton PB (2010) Mercury-organic matter relationship in pre-pollution sediments of thermokarst lakes from the Mackenzie River Delta, Canada: the role of depositional environment. Biogeochemistry. doi:10.1007/s10533-010-9543-1

Schmidt MWI, Noack NG (2000) Black carbon in soils and sediments: analysis, distributions, implications, and current challenges. Global Biogeochem Cycle 14:777–793

Selin NE, Jacob DJ, Yantosca RM, Strode S, Jaeglé L, Sunderland EM (2008) Global 3-D land-ocean-atmosphere model for mercury: present-day versus preindustrial cycles and anthropogenic enrichment factors for deposition. Global Biogeochem Cycles 22:GB2011. doi:10.1029/2007GB003040

Shi J, Ip CCM, Zhang G, Jiang G, Li X (2010) Mercury profiles in sediments of the Pearl River Estuary and the surrounding coastal area of South China. Environ Pollut 158:1974–1979

Shrestha G, Traina SJ, Swanston CW (2010) Black carbon's properties and role in the environment: a comprehensive review. Sustainability 2:294–320

Slemr F, Langer E (1992) Increase in global atmospheric concentrations of mercury inferred from measurements over the Atlantic Ocean. Nature 355:434–437

Stern GA, Sanei H, Roach P, Delaronde J, Outridge PM (2009) Historical interrelated variations of mercury and aquatic organic matter in lake sediment cores from a subarctic lake in

Yukon, Canada: further evidence toward the algal-mercury scavenging hypothesis. Environ Sci Technol 43:7684–7690
Streets DG, Gupta S, Waldhoff ST, Wang MQ, Bond TC, Bo Y (2001) Black carbon emissions in China. Atmos Environ 35:4281–4296
Streets DG, Bond TC, Carmichael GR, Fernandes SD, Fu Q, He D, Klimont Z, Nelson SM, Tsai NY, Wang MQ, Woo JH, Yarber KF (2003) An inventory of gaseous and primary aerosol emissions in Asia in the year 2000. J Geophys Res 108(D21):8809–8832
Streets DG, Hao JM, Wu Y, Jiang JK, Chan M, Tian HZ, Feng XB (2005) Anthropogenic mercury emissions in China. Atmos Environ 39:7789–7806
Streets DG, Zhang Q, Wu Y (2009) Projections of global mercury emissions in 2050. Environ Sci Technol 43:2983–2988
Sun YL, Sun M, Wei GJ, Lee T, Nie BF, Yu ZW (2004) Strontium contents of a Porites coral from Xisha Island, South China Sea: a proxy for sea-surface temperature of the 20th century. Paleoceanography 19. doi:10.1029/2003PA000959
Sun LG, Zhao SP, Liu XD, Xie ZQ, Yin XB, Liu KX, Wu XH (2005) An eco-environmental report on Xisha Archipelago, South China Sea. Chin J Nat 27(2):79–84 (in Chinese with English abstract)
Sun LG, Yin XB, Liu XD, Zhu RB, Xie ZQ, Wang YH (2006) A 2000-year record of mercury and ancient civilizations in seal hairs from King George Island, West Antarctica. Sci Total Environ 368:236–247
Sun XS, Peng P, Song JZ, Zhang G, Hu JF (2008) Sedimentary record of black carbon in the Pearl River estuary and adjacent northern South China Sea. Appl Geochem 23:3464–3472
Szopka K, Karczewska A, Kabata C (2011) Mercury accumulation in the surface layers of mountain soils: a case study from the Karkonosze Mountains, Poland. Chemosphere. doi:10.1016/j.chemosphere.2011.01.049
Tian HZ, Wang Y, Xue ZG, Cheng K, Qu YP, Chai FH, Hao JM (2010) Trend and characteristics of atmospheric emissions of Hg, As, and Se from coal combustion in China, 1980–2007. Atmos Chem Phys 10:11905–11919
UNEP Chemicals Branch (2008) The global atmospheric mercury assessment: sources, emissions and transport. UNEP-Chemicals, Geneva
Verardo DJ, Ruddiman WF (1996) Late Pleistocene charcoal in tropical Atlantic deep-sea sediments: climatic and geochemical significance. Geology 24(9):855–857
Wade E, Mizutani H, Minagawa M (1991) The use of stable isotopes for food web analysis. Crit Rev Food Sci Nutr 30(4):361–371
Wang XC, Li AC (2007) Preservation of black carbon in the shelf sediments of the East China Sea. Chinese Sci Bull 52(22):3155–3161
Wang X, Yang H, Gong P, Zhao X, Wu G, Turner S, Yao T (2010) One century sedimentary records of polycyclic aromatic hydrocarbons, mercury and trace elements in the Qinghai Lake, Tibetan Plateau. Environ Pollut 158:3065–3070
Wong CSC, Duzgoren-Aydin NS, Aydin A, Wong MH (2006) Sources and trends of environmental mercury emissions in Asia. Sci Total Environ 368:649–662
Wu YH, Wang SM, Xia WL, Liu J (2005) Dating recent lake sediments using spheroidal carbonaceous particle (SCP). Chinese Sci. Bull. 50:1016–1020
Wu Y, Wang SX, Streets DG (2006) Trends in anthropogenic mercury emissions in China from 1995 to 2003. Environ Sci Technol 40:5312–5318
Wu YH, Jiang XZ, Liu EF, Yao SC, Zhu YX, Sun ZB (2008) The enrichment characteristics of mercury in the sediments of Dongjiu and Xijiu, taihu Lake catchment, in the past century. Sci China, Ser D Earth Sci 51(6):848–854
Xie ZQ, Sun LG, Zhang PF, Zhao SP, Yin XB, Liu XD, Cheng BB (2005) Preliminary geochemical evidence of groundwater contamination in coral islands of Xi-Sha, South China Sea. Appl Geochem 20(10):1848–1856
Xu B, Yang X, Gu Z, Zhang Y, Chen Y, Lv Y (2009) The trend and extent of heavy metal accumulation over last one hundred years in the Liaodong Bay, China. Chemosphere 75:442–446

Xu LQ, Liu XD, Sun LG, Yan H, Liu Y, Luo YH, Huang J, Wang YH (2010) Distribution of radionuclides in the guano sediments of Xisha Islands, South China Sea and its implication. J. Environ. Radioactiv. 101(5):362–368

Xu LQ, Liu XD, Sun LG, Yan H, Liu Y, Luo YH, Huang J (2011a) Geochemical evidence for the development of coral island ecosystem in the Xisha Archipelago of South China Sea from four ornithogenic sediment profiles. Chem Geol 286:135–145

Xu LQ, Liu XD, Sun LG, Chen QQ, Yan H, Liu Y, Luo YH, Huang J, Wang YH (2011b) A 700-year record of mercury in avian eggshells of Guangjin Island, South China Sea. Environ Pollut 159:889–896

Yan H, Sun LG, Wang YH, Liu XD, Qiu SC, Cheng WH (2010) A 2000-year record of copper pollution in South China Sea derived from seabird excrements: a potential indicator for copper production and civilization of China. J Paleolimnol 44:431–442

Yang H, Rose N (2003) Distribution of mercury in six lake sediment cores across the UK. Sci Total Environ 304:391–404

Yang HD, Battarbee RW, Turner SD, Rose NL, Derwent RG, Wu G, Yang R (2010a) Historical reconstruction of mercury pollution across the Tibetan Plateau using lake sediments. Environ Sci Technol 44:2918–2924

Yang HD, Engstrom DR, Rose NL (2010b) Recent changes in atmospheric mercury deposition recorded in the sediments of remote equatorial lakes in the Rwenzori Mountains, Uganda. Environ Sci Technol 44:6570–6575

Yin XB, Xia LJ, Sun LG, Luo HH, Wang YH (2008) Animal excrement: a potential biomonitor of heavy metal contamination in the marine environment. Sci Total Environ 399:179–185

Zhao HT (1996) History of expeditions to Xisha Islands. Geographical Research 15(4):55–65 (in Chinese with English abstract)

# Chapter 9
# A Preliminary Study of Ancient DNA in Guano Subfossils from the Xisha Islands

## 9.1 Climate Change and Evolution

Secret of life, including its origin and evolution, is one of the most mysterious things of the nature. Scientists have make lots of efforts to find the answer. Climate changes can significantly impacts on macro-traits of creatures, i.e. population size, foraging strategy and reproductive behavior (Sun et al. 2000; Emslie and Patterson 2007; Huang et al. 2011). At the same time, abrupt change in climate may also impose strong selection pressures on traits crucial for fitness, and therefore, microevolution is potentially an important mechanism mitigating adverse effects of climate oscillation (Gienapp et al. 2008). For example, Lambert et al. (2002) estimated evolution rates in Ancient DNA from penguin bones, and reported that evolution rate over the last 7,000 years was two to seven times higher than phylogenetic estimates. A recent study of North American mammals showed that diversity pattern of faunas mirrored stacked benthic foraminiferal $\delta^{18}O$ record, a marker for global ice volume, implying a remarkable impact of climate on ancient terrestrial mammals over the Cenozoic (Figueirido et al. 2012). In addition to climate change, extensive and significant human forcing can impose evident threat to megafauna as well. The extinction of terrestrial megafauna during the late Quaternary is a widespread event, but its causes are still highly controversial. Prescott et al. (2012) probed the possible reasons by examining the timing and distribution of widespread megafaunal extinctions on five continents, and suggested that both climate dynamics and human ingredient were involved in the extinctions of megafauna, including ground sloths, mammoths, cats etc., in the late Quaternary. Via phylogeographic, morphometric and ancient DNA analysis of both contemporary and ancient deer remains (dating back to 30,000–1,700 years), Carden et al. (2012) emphasized a vital role of Neolithic people on fauna and flora in Ireland and Britain.



L. Xu, *Impact of Climate Change and Human Activity on the Eco-environment*, Springer Theses, DOI 10.1007/978-3-662-45003-1_9

Understanding the impact of climate on the ecology and evolution of organisms is of great significance for preventing future biodiversity collapse. However, the change of genetic structure, i.e. DNA, in response to climate change and anthropogenic activity has not been well documented so far, and explicit evidence of adaptation to ongoing warming remains scare. The study of ancient DNA from nucleic acid-bearing biological materials is essential to examine the possible interplay among climate change, evolution and human activity.

## 9.2 Ancient DNA

### *9.2.1 Ecological Implications of Ancient DNA from a Variety of Biological Materials*

Ancient DNA provides important information about lives in the past, and the study of ancient DNA has drawn increasing attention in recent years. DNA extraction is an essential step in such studies. Isolation of DNA from ancient samples, however, is not as easy as that from living organisms. With technical advances in DNA isolation, scientists have successfully isolated DNA from a variety of well-preserved biological remains to study paleontology and palaeoecology. As given in Sect. 9.1, Lambert et al. (2002) successfully isolated ancient DNA from subfossil bone of penguins dating approximately 7,000 years ago. Biological remains of extinct bird giant moa from New Zealand have been widely used in ancient DNA study. For example, using fossil bone samples, Baker et al. (2005) studied ancient mitochondria DNA in extinct giant moa remains from New Zealand and found lineage-splitting occurred approximately $10^{-4}$ million years ago with general climate cooling. This coincided with landmass split driven by tectonic and mountain-building movements during that epoch. It is known that feathers contain amplifiable DNA at their calamus. Amplification of DNA from upper portion of feathers is normally quite difficult due to its low yield. Despite of the low success probability, Rawlence et al. (2009) isolated sufficient ancient DNA from Holocene New Zealand moa feathers, allowing a means to reconstruct appearance, and species identification of extinct avian taxa. Following these studies, Oskam et al. (2010) successfully isolated and amplified ancient DNA fragments from 19 ka years old fossil eggshells for the first time, suggesting that eggshell remains were previously underestimated sources of ancient DNA.

Of those biological remains, excrement is a special material used for DNA study; it contains DNA from not only host organism, but also the organism's diets. As droppings are waste products, the analysis of prey DNA in faeces thus acts as a non-invasive approach to investigate the diets of organisms, including birds. They have been used to reconstruct diets of both herbivores and carnivores. For example, Valentini et al. (2009a) examined the feasibility of isolating DNA fragments, i.e. chloroplast *trn*L intron, from wild herbivore faeces. They reported that short size DNA segments (<100–150 bp) appeared in a higher proportion in degraded DNA, and can be successfully recovered. Poinar et al. (1998) used a chemical reagent, *N*-phenacylthiazolium bromide which can cleave DNA cross-link and further enhance amplification efficiency. They isolated DNA from unknown ancient coprolite samples and identified they were derived from extinct sloths. According to prey DNA analysis by little penguin faeces, Deagle et al. (2010) found that proportions of prey considerably differed from mass proportions in the diet. They thus argued that dietary DNA analysis could serve as a semi-quantitative approach.

### *9.2.2 Technical Problems in Ancient DNA Study*

Ancient DNA contains information pertinent to various study fields, including evolution, ecology, climatology, archaeology etc., and thus creates a new perspective on DNA sequences, evolution of creatures and environmental changes. However, this field is filled with technical pitfalls and needs more strict criteria to ensure the reliability of experimental results (Hofreiter et al. 2001). The major obstacles of ancient DNA study are its pretty low yield, contamination with contemporary DNA and extensive degradation (Mitchell et al. 2005). The problems in the study of ancient DNA are summarized as two types, which are as follows.

1. Contamination with exogenous DNA molecular

   DNA contamination is a pervasive problem, especially for ancient DNA. For example, genetic analysis of 24 Neandertal DNA yielded DNA sequences identical or similar to those of modern humans (Serre et al. 2004). Pääbo (2004) attributed this to a potential DNA contamination of contemporary human DNA. DNA cloning from a quagga (Higuchi et al. 1984) and an Egyptian mummy (Pääbo 1985) may also be caused by introduction of exogenous DNA, as reproducibility was impossible to achieve in those experiments Pääbo (2004). To avoid contamination with exogenous DNA, scientists initialized standard procedures to improve ancient DNA study. For example, extraction/PCR controls and biochemical analysis of macromolecular preservation need to be involved in the study of ancient DNA (Pääbo 2004). As ancient DNA can be readily contaminated, field sample collection and following experiments have to be with special care.

2. Molecular damage

   Another troublesome problem in ancient DNA study is its structure changes, rendering it not as the same as its original formation. These processes involve strand breaks, oxidative lesions, DNA cross-link, hydrolytic lesions etc. The most obvious type of damages to ancient DNA from subfossil/fossil remains is its degradation to small sizes, normally <500 bp (Deagle et al. 2006). Studies have shown that only short segments of DNA can be amplified from ancient remains (Pääbo 2004). Another type of DNA damage is cross-link, which block the DNA polymerase. Hydrolytic loss of amino groups is a third common form of DNA damages, and this may lead to incorporation of incorrect bases during amplification.

## 9.3 Isolation of DNA from Guano Samples of the Xisha Islands

### *9.3.1 DNA Barcoding*

DNA barcoding is a molecular biological approach for species identification via sequence analysis of standardized DNA regions. The concept of DNA barcoding was firstly defined in a widely cited paper namely "Biological identifications through DNA barcodes" written by Professor Paul D.N. Hebert's research group in 2003 (Hebert et al. 2003). Since then, biologists have made extensive use of this technique for taxon identification. For animals, cytochrome c oxidase 1 (COI) is an ideal barcode. The situation for plants, however, is controversial. Barcodes for plants based on either a chloroplast DNA fragment, or a combination of several regions (Valentini et al. 2009b). For example, Hebert et al. (2004) determined COI barcodes of 260 species of North American birds and identified four new species using species-specific COI barcodes. Yoo et al. (2006) reported that COI barcode is quite effective in distinguishing more than 200 Korean bird species. DNA barcoding also works well for fish species (Ward et al. 2005; Ivanova et al. 2007). Recently, Zhang and Hanner (2012) applied DNA barcoding to identification of fish in the South China Sea, and suggested that mitochondrial 16S rRNA (16S) and nuclear ribosomal 18S rRNA (18S) also have the ability to identify fish species. Scientists have now well-established DNA barcode database, such as Consortium for the Barcode of Life (CBOL), Barcode of Life Data Systems (BOLD), International Barcode of Life (iBOL) etc.

DNA barcoding has been widely used in ecological study. Figure 9.1 shows the methodology for analyzing biodiversity from a variety of environmental samples based on DNA barcoding. There are lots of guano particles from the ornithogenic sediments of the Xisha Islands. Due to a relatively high temperature and poorly developed groundwater system, these guano samples have been physically well-preserved and they may contain ancient DNA. Our tentative idea is to examine possible responses of bird/fish genes to climate change and human activity, via bird and fish DNA barcoding (Fig. 9.2).

### *9.3.2 Isolation of DNA from Ancient Guano Samples*

DNA from biological materials can be isolated by means of alkali extraction or commercial DNA isolation kit. In general, DNA isolated by commercial kit is with high purity, but has a low yield. In contrast, alkali extraction yields a relatively high concentration DNA, but its quality is low.

1. Alkali extraction

   Alkali extraction of DNA is a traditional means and has been widely employed. The design of DNA isolation from ancient guano is given in Fig. 9.3. This

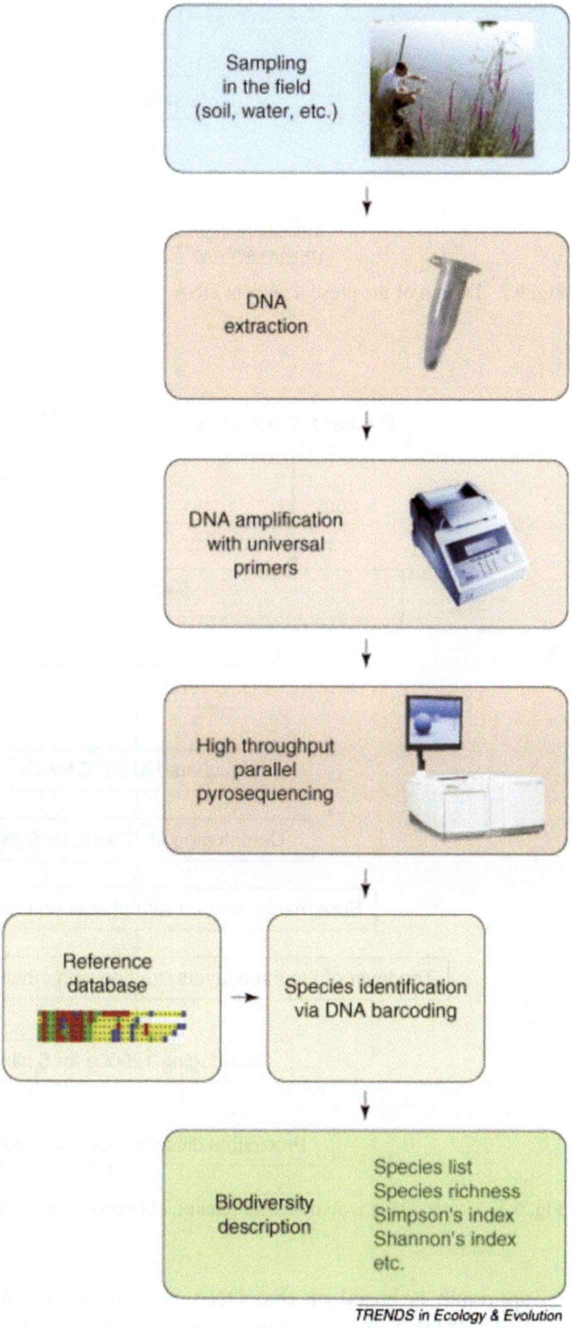


**Fig. 9.1** Methodology for DNA barcoding (adapted from Valentini et al. 2009b). [Reprinted from Valentini et al. 2009b, Copyright (2009), with permission from Elsevier.]

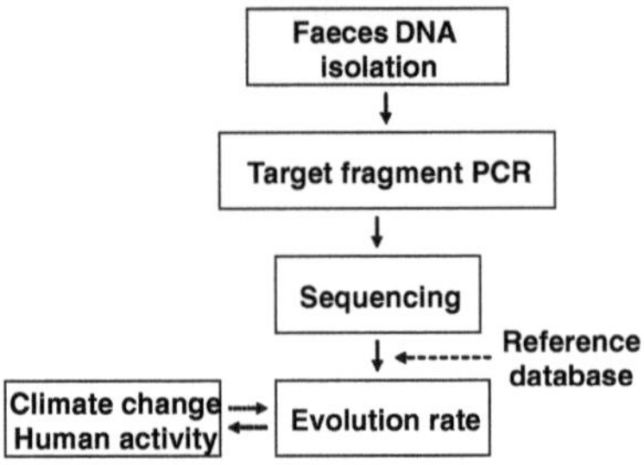


**Fig. 9.2** Design of the present ancient DNA study

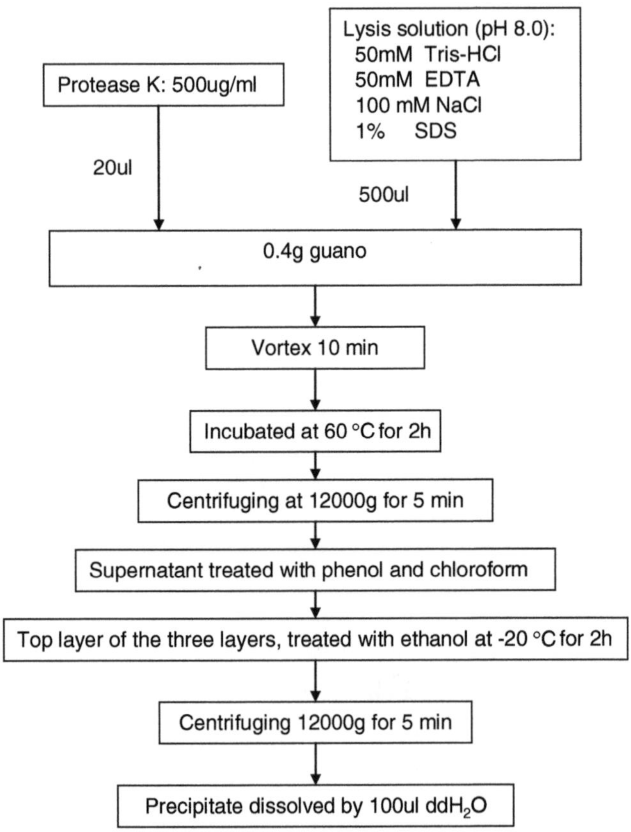


**Fig. 9.3** Guano DNA isolation with phenol, chloroform and ethanol

approach is based on that DNA is soluble in water, but not in alcohols. The isolated DNA from ancient guano samples are shown as bands 1–5 in Fig. 9.4. From this figure, it is evident that the guano from the Xisha Islands contains ancient DNA. Furthermore, we can also conclude that the more ancient the sample is, the lower the yield is. It seems that yield is highly related with time.

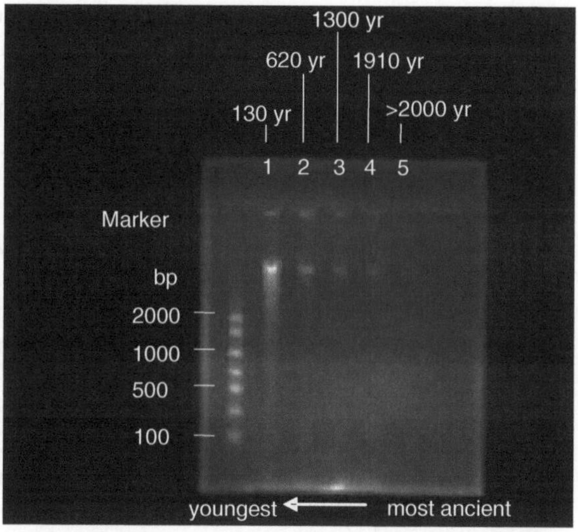


**Fig. 9.4** Ancient DNA from ancient guano samples via alkali extraction

If ancient DNA degraded at a relatively constant rate, it may be used to estimate ages of DNA-bearing samples.

The problem of alkali extraction is that quality of the isolated DNA samples is too low and not suitable for following analysis. To isolate high quality DNA, We also tested the feasibility of using commercial kit to isolate ancient DNA from guano samples.

2. Commercial kit

   Following the standard procedure specified in the kit (QIAGEN Stool DNA isolation kit) protocol, we attempted to isolate any genetic fragment from ancient guano samples. However, we did not observe any band in the running gel under ultraviolet. This may be explained by both loss during isolation and pretty low level of ancient DNA in guano due to severe degradation. Polymerase China Reaction (PCR) is a technique that makes it possible to produce large number of DNA fragment copies from the isolated DNA template.

### *9.3.3 Polymerase Chain Reaction (PCR)*

PCR is a widely used technique in molecular biology to amplify DNA pieces across several orders of magnitude along with a DNA polymerase. Short DNA fragments, those are complementary to target region in DNA template, are called primers. To amplify bird COI gene, primers BirdF1 (5′–3′ sequence TTCTCCAACCACAAAGACATTGGCAC) and BirdR1 (5′–3′ sequence ACGTGGGAGATAATTCCAAATCCTG) have been proven to be effective (Hebert et al. 2004).

To optimize PCR conditions, DNA isolated from fresh chicken faeces was used before we tested guano DNA. Genomic DNA was extracted by QIAGEN Stool DNA isolation kit. DNA amplification was conducted in a final volume of 25 μL. The 25 μL reaction mixes included 12.5 μL KAPA2G Robust HotStart ReadyMix (2X), 1 μL DNA template, 1.25 μL each primer (BirdF1/BirdR1, 10 μM), 1.25 μL DMSO and 2.5 μg bovine serum albumin (BSA). The amplification regime consisted of 2 min at 94 °C followed by 35 cycles of 1 min at 94 °C, 1.5 min at an annealing temperature and 1.5 min at 72 °C, and a final 5 min 72 °C. The PCR products were then visualized in a 1.5 % agarose gel.

The original annealing temperature specified in the standard protocol of KAPA2G Robust HotStart ReadyMix is 60 °C. When we tried this annealing temperature, none of PCR products were visualized in the agarose gel under ultraviolet (Fig. 9.5a). Then the annealing temperature was changed to 51 °C. The PCR products were detected, but they appeared as smears (Fig. 9.5b). BirdF1 was then substituted by an alternative reverse primer BirdR3 (5′–3′ sequence: AGGAGTTTGCTAGTACGATGCC, Hebert et al. 2004). The gel picture was shown in Fig. 9.5c, from which we observed non-specific products. Spiess et al. (2004) reported that trehalose could be a good PCR enhancer. We added this reagent to the PCR mix (final concentration of 5 %), and got a better result (Fig. 9.5d). Although there remained a non-specific band, the target product was clear.

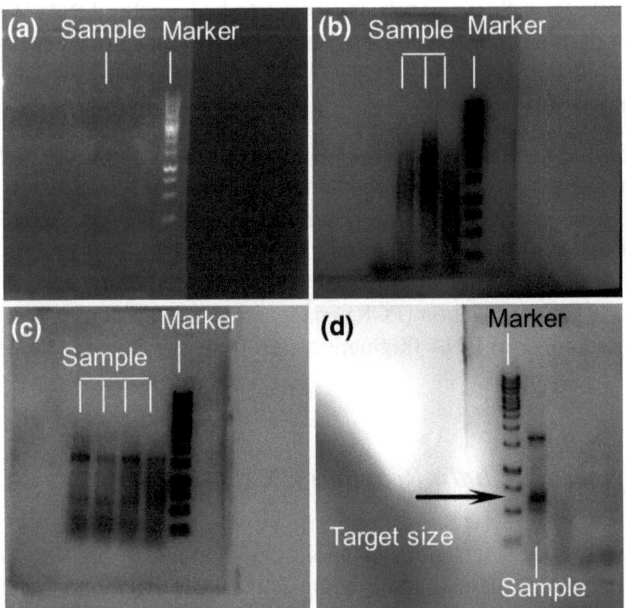


**Fig. 9.5** Amplification of COI gene from fresh chicken faeces. **a** Annealing temperature 60 °C, primers-BirdF1/BirdR1, no PCR product; **b** annealing temperature 51 °C, primers-BirdF1/BirdR1, smears appeared; **c** annealing temperature 51 °C, primers-BirdF1/BirdR3, non-specific products; **d** annealing temperature 51 °C, primers-BirdF1/BirdR2, Trehalose added

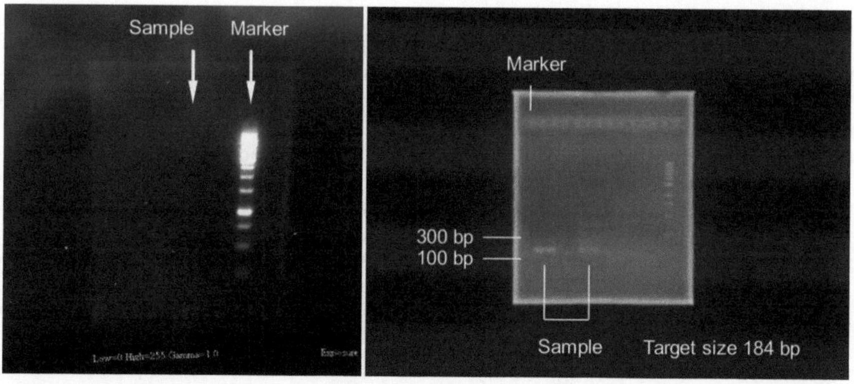


**Fig. 9.6** Amplification of COI gene from a three-year old guano sample. *Left* primer set BirdF1/BirdR3; *Right* mini-barcode primer set BirdF1/GE01, for a three-year old sample (concentration of the marker in the *right figure* is very low)

We isolated DNA from a three-year old guano sample by QIAGEN Stool DNA isolation kit. Using optimized PCR conditions, it was attempted to amplify COI gene from it. However, we failed to amplify this gene (see left panel in Fig. 9.6). This is probably attributed to severe degradation of faeces DNA into pieces. Scientists have found that short baocodes (so called mini-barcode) can also identify a specimen, whose DNA is severely degraded (Hajibabaei et al. 2006; Rudnick et al. 2007). Primers BirdF1 and GE01 (5′–3′ sequence: CTATGAAGAAGATTATTACRAAAGCA) amplify a short bird DNA COI gene (184 bp, Rudnick et al. 2007). It seems that this primer set worked well for the three-year old guano sample (Fig. 9.6, right panel). Our next step is trying to amplify mini-barcodes from more ancient DNA from guano samples.

From DNA isolation from ancient guano samples and following PCR, we conclude that ancient guano samples contained DNA, although they have severely degraded. Nonetheless, they have a potential use to amplify bird COI genes.

## References

Baker AJ, Huynen LJ, Haddrath O, Millar CD, Lambert DM (2005) Reconstructing the tempo and mode of evolution in an extinct clade of birds with ancient DNA: the giant moas of New Zealand. Proc Natl Acad Sci USA 102(23):8257–8262

Carden RF, McDevitt AD, Zachos FE, Woodman PC, O'Toole P, Rose H, Monaghan NT, Campana MG, Bradley DG, Edwards CJ (2012) Phylogeographic, ancient DNA, fossil and morphometric analyses reveal ancient and modern introductions of a large mammal: the complex case of red deer (Cervus elaphus) in Ireland. Quat Sci Rev. doi:10.1016/j.quascirew.2012.02.012

Deagle BE, Eveson JP, Jarman SN (2006) Quantification of damage in DNA recovered from highly degraded samples—a case study on DNA in faeces. Front Zool 3:11

Deagle BE, Chiaradia A, McInnes J, Jarman SN (2010) Pyrosequencing faecal DNA to determine diet of little penguins: is what goes in what comes out? Conserv Genet 11:2039–2048

Emslie SD, Patterson WP (2007) Abrupt recent shift in $\delta^{13}C$ and $\delta^{15}N$ values in Adelie penguin eggshell in Antarctica. Proc Natl Acad Sci 104:10213–10217

Figueirido B, Janis CM, Perez-Claros JA, Renzi MD, Palmqvist P (2012) Cenozoic climate change influences mammalian evolutionary dynamics. Proc Natl Acad Sci USA 109(3):722–727

Gienapp P, Teplitsky C, Alho JS, Mills JA, Merila J (2008) Climate change and evolution: disentangling environmental and genetic responses. Mol Ecol 17:167–178

Hajibabaei M, Smith MA, Janzen DH, Rodriguez JJ, Whitfield JB, Hebert PDN (2006) A minimalist barcode can identify a specimen whose DNA is degraded. Mol Ecol Notes 6:959–964

Hebert PDN, Cywinska A, Ball SL, deWaard JR (2003) Biological identifications through DNA barcodes. Proc R Soc Lond B 270:313–321

Hebert PDN, Stoeckle MY, Zemlak TS, Francis CM (2004) Identification of birds through DNA barcodes. PLoS Biol 2(10):e312

Higuchi R, Bowman B, Freiberger M, Ryder OA, Wilson AC (1984) DNA sequences from the quagga, an extinct member of the horse family. Nature 312:282–284

Hofreiter M, Serre D, Poinar HN, Kuch M, Pääbo S (2001) Ancient DNA. Nature 2:353–359

Huang T, Sun LG, Stark J, Wang Y, Cheng ZQ, Yang QC, Sun S (2011) Relative changes in krill abundance from Antarctic fur seal. PLoS ONE 6(11):e27331. doi:10.1371/journal.pone.0027331

Ivanova NV, Zemlak TS, Hanner RH, Hebert PDN (2007) Universal primer cocktails for fish DNA barcoding. Mol Ecol Notes 7(4):544–548

Lambert DM, Ritchie PA, Millar CD, Holland B, Drummond AJ, Baroni C (2002) Rates of evolution in ancient DNA from Adelie Penguins. Science 295:2270–2273

Mitchell D, Willerslev E, Hansen A (2005) Damage and repair of ancient DNA. Mutat Res 571:265–276

Oskam CL, Haile J, McLay E, Rigby P, Allentoft ME, Olsen ME, Bengtsson C, Miller GH, Schwenninger J, Jacomb C, Walter R, Baynes A, Dortch J, Parker-Pearson M, Gilbert MTP (2010) Fossil avian eggshell preserves ancient DNA. Proc R Soc B 277:1991–2000

Pääbo S (1985) Molecular cloning of ancient Egyptian mummy DNA. Nature 314:644–645

Pääbo S (2004) Genetic analyses from ancient DNA. Ann Rev Genet 38:647–679

Poinar HN, Hofreiter M, Spaulding WG, Martin PS, Stankiewicz BA, Bland H, Evershed RP, Possnert G, Pääbo S (1998) Molecular coproscopy: dung and diet of the extinct ground sloth Nothrotheriops shatensis. Science 281:402–406

Prescott GW, Williams DR, Balmford A, Green RE, Manica A (2012) Quantitative global analysis of the role of climate and people in explaining late Quaternary megafaunal extinctions. Proc Nat Acad Sci USA 109(12):4527–4531

Rawlence NJ, Wood JR, Armstrong KN, Cooper A (2009) DNA content and distribution in ancient feathers and potential to reconstruct the plumage of extinct avian taxa. Proc R Soc B 276:3395–3402

Rudnick JA, Katzner TE, Bragin EA, Dewoody JA (2007) Species identification of birds through genetic analysis of naturally shed feathers. Mol Ecol Notes 7:757–762

Serre D, Langaney A, Chech M, Teschler-Nicola M, Paunovic M, Mennecier P, Hofreiter M, Possnert G, Pääbo S (2004) Neandertal mtDNA contribution to early modern humans. PLoS Biol 2(3):e57

Spiess A, Mueller N, Ivell R (2004) Trehalose is a potent PCR enhance: lowering of DNA melting temperature and thermal stabilization of Taq Polymerase by the disaccharide trehalose. Clin Chem 50:1256–1257

Sun LG, Xie ZQ, Zhao JL (2000) A 3000-year record of penguin population. Nature 407:808

Valentini A, Miquel C, Nawaz MA, Bellemain E, Coissac E, Pompanon F, Gielly L, Cruaud C, Nascetti G, Wincker P, Swenson JE, Taberlet P (2009a) New perspectives in diet analysis based on DNA barcoding and parallel pyrosequencing: the *trn*L approach. Mol Ecol Resour 9:51–60

Valentini A, Pompanon F, Taberlet P (2009b) DNA barcoding for ecologists. Trends Ecol Evol 24(2):110–117

Ward RD, Zemlak TS, Innes BH, Last PR, Hebert PDN (2005) DNA barcoding Australia's fish species. Phil Trans R Soc B 360:1847–1857

Yoo HS, Eah JY, Kim JS, Kim YJ, Min MS, Oaek WK, Lee H, Kim CB (2006) DNA barcoding Korean birds. Mol Cells 22(3):323–327

Zhang JB, Hanner R (2012) Molecular approach to the identification of fish in the South China Sea. PLoS ONE 7(2):e300621

# Chapter 10
# Conclusions

Based on multi-analysis of bird remains and bulk sediments from Ganquan, Guangjin, Jinqing, Jinyin and Chenhang islands of the Xisha archipelago, it is concluded that:

1. The $^{210}Pb$ dating technique is applicable for the widely distributed ornithogenic sediments in the Xisha Islands. The mean atmospheric $^{210}Pb$ supply rate (or flux) in the Xisha Islands is very close to the average $^{210}Pb$ flux of northern hemisphere. Human nuclear tests did not have a significant impact on the Xisha Islands. The main source of radionuclides in the ornithogenic coral sand sediments is atmospheric precipitation, and the organic matter derived from plant and guano could further enhance them. Radiocarbon analysis of bone samples in the ornithogenic sediments suggests that these sediments are genuine natural archive and have been well-preserved. Those sediments were deposited within approximately the last 2,000 years and thus provide information about seabird ecology over the past two millennia.
2. The source of sedimentary materials abruptly changed at a critical depth (11, 16, 9, and 12 cm in GQ, GJ3, JQ and JY2, respectively), revealing that the sedimentary components have changed gradually from a two-component (coral sand and guano) system to a three-component mixture (coral sand, guano and humus). This implies a lag in plant development relative to seabird inhabitation and may be the typical mode of ecological development in the coral islets of the Xisha Islands, South China Sea. Cu, Cd, Zn, P, As, Se and Ba were identified as avian bio-elements, and could be used to track seabird population.
3. The reconstructed seabird population suggested that seabirds began to inhabit Ganquan Island more than 2,200 years ago. Seabird population over the past 2,200 years can be divided into several stages. As there were more seabirds during the cold LIA, it seems that a cool climate is more favorable to seabirds on the Xisha Islands. Low SST and strong East Asian winter monsoon during cool times enhanced seawater and nutrient mixing in the ocean, which further



L. Xu, *Impact of Climate Change and Human Activity on the Eco-environment*,
Springer Theses, DOI 10.1007/978-3-662-45003-1_10

facilitated marine primary production and food availability for seabirds. This ultimately supported a large size of seabird populations.

4. Isotope-based diets of seabirds suggest significant changes in seabird foraging strategy over the past 2,000 years. Seabird population on the Xisha Islands over the past 2,000 years was closely and positively related to its trophic position: more seabirds corresponded to higher collagen $\delta^{15}N$, or trophic level. This may be attributed to competitions among birds. When there are more seabirds, a fierce intra/inter-specific competition may force them to dive deeper, or migrate from inshore habitats to offshore areas, and feed on preys with higher tropic level.
5. The 700-year records of eggshell Hg and atmospheric Hg flux well-recorded past human metallurgical activity, including enhanced metallurgy activity in Ming Dynasty of China, exploitation on the Europe/America and World War II etc. They also reflect a gradual shift of Hg production center from Europe and America to Asia, since 1970s. A 400-year record of black carbon (BC) has been reconstructed. The BC flux in the Xisha Islands was low before the onset of 20th century, it had a rapid increase since the middle of twentieth century, and it peaked in 1970s. The gradual decrease of BC in the past 30 years possibly reflected the change of energy structure from biomass to fossil fuels and the improvement of pollution control technique in the regions around the South China Sea.
6. Although ancient DNA has severely degraded, DNA fragments can also be isolated from ancient guano samples. The yield, however, is quite low.
7. Seabird population has significantly decreased in recent times, and they have disappeared on most of the small islets of the Xisha Islands. Heavy metal pollution, may pose a threat to the fragile coral island ecosystem. We thus strongly advocate a broad focus on and open discussion in the management of tropical atolls.

MIX
Papier aus verantwortungsvollen Quellen
Paper from responsible sources
FSC® C105338

If you have any concerns about our products,
you can contact us on
**ProductSafety@springernature.com**

In case Publisher is established outside the EU,
the EU authorized representative is:
**Springer Nature Customer Service Center GmbH**
**Europaplatz 3, 69115 Heidelberg, Germany**

Printed by Libri Plureos GmbH
in Hamburg, Germany